D0221949

The RIGHT ANTENNA

How to Select and Install Antennas for Entertainment and Communications Devices

By
Alvis J. Evans

With Contributions By:
Jerry Luecke

PROMPT™
PUBLICATIONS

REVISED SECOND EDITION, 1992

PROMPT® Publications is an imprint of Howard W. Sams & Company , 2647 Waterfront Parkway, East Drive Indianapolis, IN 46214-2041.

This book was originally developed and then published as Antennas *by:*
        Master Publishing, Inc.
        14 Canyon Creek Village MS31
        Richardson, Texas 75080
        (214) 907-8938

International Standard Book Number: 0-7906-1022-1

Edited by: *Roger R. Webster, Fellow, Institute of Electrical & Electronic Engineers*
Test Design and Artwork By: *Plunk Design, Dallas, TX*
Cover Design by: *Sara Wright*

Acknowledgements
All photographs not credited are either courtesy of Author, Master Publishing, Inc., or Howard W. Sams & Company.

Printed in the United states of America

Table of Contents

Preface

In this age of information, each of us is affected by the advances in electronic communications such as AM/FM radio, television and mobile telephone. In these various modes of communications, the information travels from the transmitter to the receiver through space by electromagnetic waves. Antennas are required at both ends for the purpose of coupling the transmitter and the receiver to the space link between them. *The Right Antenna* is intended to provide easy to understand information on a wide variety of antennas. Early explanation of the basic concepts of how any antenna works helps the reader understand the later explanations of the specialized types of antennas. With this understanding, readers will be better prepared to select the best antenna for their needs.

The Right Antenna begins by explaining how antennas work in general and then isolates antennas for TV and FM. A separate chapter is devoted to satellite TV antennas, TV and FM noise and interference, CB and cellular antennas, and antennas used by "hams" for amateur band operation. The basic concepts of cellular telephone system operation are explained and the most popular antennas are discussed. The chapter on the fast-growing area of satellite TV provides enough information for readers to intelligently select their own satellite system.

The antenna problems encountered in fringe areas, high noise areas, and with various types of interference are discussed, and practical solutions offered. In order for any antenna to be effective, it must be the right type for the job, and be safely and properly installed. For that reason, the book concludes with a chapter on antenna installation. A safety checklist is included.

The Right Antenna is rather broad in its scope of coverage, though we have attempted to include sufficient depth of coverage to provide a handy reference to the student, technician, or consumer. After reading this book you should be able to select an antenna, place it correctly, and install it properly to obtain maximum performance whether in a strong signal area or in a fringe area. We hope we have met this goal.

AJE

How Antennas Work

TV antennas mounted on roofs, attached to chimneys, and mounted in attics; and large metal towers used by radio stations (most with straight mast antennas and blinking warning lights) are common types of antennas that are easily recognized. The purpose of an antenna is to transmit or receive radio frequency energy. The function of an antenna when used at a transmitter is to convert the energy generated by a transmitter into a radiated electromagnetic wave. When used at the receiver, the function of an antenna is to convert the radiated wave into useful radio frequency energy for the receiver. The radiated electromagnetic waves, when intercepted by the receiving antenna, induce a voltage along the length of the antenna. The magnitude of the induced voltage depends on the strength (or intensity) of the radiated wave at the receiving antenna.

In this first chapter, we will examine the fundamental principles of radio waves, and the basic operating action and characteristics of antennas.

RADIO WAVES

Radio waves transmitted into space are radiant energy, similar to heat or light. In free space, they travel at the speed of light — 186,000 miles per second (300,000,000 meters per second). In any other medium, such as air, water, or a transmission line (a pair of wires, coaxial cable, etc.), they travel at reduced speeds. Radiated radio frequency energy from the transmitting antenna moves through space as a moving-field wave. We will look more at this wave shortly, but first let's look at how the waves reach the receiving antenna.

THE IONOSPHERE

Various gases — oxygen, nitrogen, hydrogen and helium — make up the atmosphere above the earth's surface. Oxygen and nitrogen are the main ingredients up to 50 miles, above that are hydrogen and helium. Ultraviolet radiation from the sun ionizes the gases at the higher altitudes, and produces positive and negative ions, and, in addition, free electrons. As shown in *Figure 1-1*, the ionized gases are distributed in layers of different ion density. The ionized layers of gas bend and reflect some of the radiated radio waves back to earth. If the

Figure 1-1. Relative Distribution of the Ionosphere Layers About the Earth

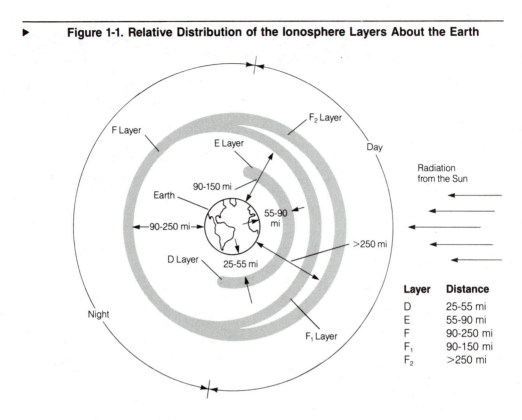

Layer	Distance
D	25-55 mi
E	55-90 mi
F	90-250 mi
F₁	90-150 mi
F₂	>250 mi

layers were not ionized, all the radiated waves would travel in a straight line, eventually leaving the earth. The path of the radiated waves in the upper atmosphere is determined by the ion density of the distributed layers. The higher the density, the greater the reflection.

Sky Waves

The moving-field waves radiated from the transmitting antenna travel into space in all directions. Waves that are radiated toward the upper atmosphere are called *sky-waves*. As stated previously, some of the sky waves are reflected back to the earth by the ionized layers in the upper atmosphere. As shown in *Figure 1-2*, the distance from the transmitter to the reflected wave's return to the earth (the *skip-distance*) can be great. Sky wave reflection normally occurs at signal frequencies from 2 MHz (2 millions cycles per second) to 30 MHz, and are commonly called "short waves."

The travel path (propagation characteristics) of sky waves are affected by two factors — *critical angle* and *critical frequency* — illustrated in *Figure 1-2*. If the radiated wave enters the ionosphere at an angle greater than θ (the critical angle), then the wave is not reflected back to the earth; while if the angle is less than the critical angle (*Figure 1-2a*), it will be bent back and eventually reach the earth again.

The height of the ionized layer will greatly affect the *skip-distance*. Skip distance also varies with the frequency of the transmitted wave. Both of these effects are shown in *Figure 1-2b*. Because the height and density of the ionized layers in the ionosphere depend on the sun's radiation, there is a significant difference between day and night transmission and skip-distance.

▶ **Figure 1-2. Radio Waves Entering Ionosphere at Various Angles and Various Frequencies**

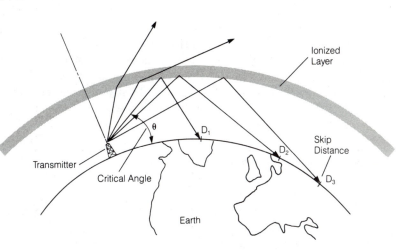

a. Effect of Various Angles on Skip Distance (All at the Same Frequency)

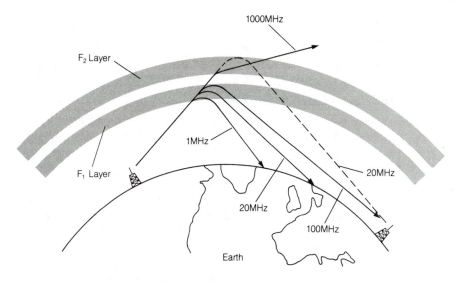

b. The Variation in Skip Distance with Frequency

Ground Waves

Ground waves, after radiation from the transmitting antenna, stay close to the earth's surface. *Figure 1-3* shows that there are three components of the ground wave — direct wave, surface wave and reflected wave. It is easy to see why a receiving antenna well below the horizon may not receive a direct wave.

Ground waves are affected by the conductivity and surface features of the earth. Higher conductivity gives better transmission, so ground waves travel best over sea water, fresh water, or flat loamy soil. Rocky terrain and desert are very poor, while jungle areas are virtually unusable.

Moisture conditions in the air near the ground will affect ground waves. A moisture inversion in the first 100 feet or so above the earth's surface channels the ground wave so that it follows the curvature of the earth for hundreds of miles. This condition sometimes exists after widespread rain or very heavy dew, and accounts for the occasional reception of TV stations hundreds of miles away. The propagation characteristics of the ground wave also are affected by the frequency of the wave.

▶   **Figure 1-3. Components of the Ground Wave**

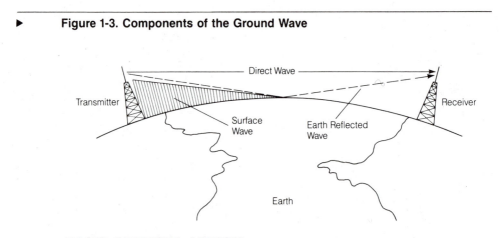

BASIC ANTENNA ACTION

Whether the antenna is receiving electromagnetic energy or transmitting it, there are two fields developed along the antenna. An electrostatic field from end to end due to a voltage along the antenna (*Figure 1-4a*), and a magnetic field around the antenna due to a current in the antenna (*Figure 1-4b*). Even though the basic operation of all antennas is the same, explaining the interaction of the two fields requires complex mathematical equations to describe the relationships. James Clerk Maxwell developed the basic concepts and the complex "Maxwell Equations" in the 1800's.

In the paragraphs which follow, we will be discussing many characteristics of antennas. It is important to realize that all of the characteristics of a given antenna (such as gain, directivity, impedance, polarization, efficiency, etc.) are the same whether the antenna is used for receiving or transmitting.

Some Basic Concepts

Any wire carrying a current has a magnetic field surrounding the wire. You can verify this with a pocket compass, a piece of wire (a bent paper clip will do) and a small (AA) flashlight cell. Place the wire near the compass, and momentarily touch the ends of the wire to the cell. The compass will jump.

It seems reasonable to assume that some energy is required to establish a magnetic field. Not so obvious is that a steady (static) magnetic field requires no energy to maintain it. However, think of a permanent magnet: once magnetized, no further energy is required to maintain the field. Further, if we stop the current in the wire, the magnetic field will collapse, trying to return the energy in the field to the wire.

How far away does a magnetic field around a wire extend? If we had a sufficiently sensitive detector, and there were no other magnetic fields to confuse us, we would find that the magnetic field extends to a very large distance — essentially infinite!

To further explore how an antenna works, let's consider induced currents. If we place two wires side by side, and connect a sensitive detector to the ends of one, while we alternately connect and disconnect our AA cell to the ends of the other, the detector will indicate whenever we EITHER connect OR disconnect the cell, but NOT while the cell is either connected or disconnected. Thus, we can induce a voltage in the second wire whenever the current in the first wire changes. This is a very important concept.

If the current in the first wire were alternating (first in one direction and then in the other) then the changing current will induce an alternating voltage in the second wire. One alternation back and forth is called a cycle and the number of cycles per second is the frequency. The name given to cycles per second is Hertz, after Heinrich Rudolph Hertz, an early German scientist who investigated electricity and discovered electromagnetic waves. If the frequency of alternation were millions of Hertz (megahertz, abbreviated MHz), we would have radio frequencies; if it alternates 60 to 200 million Hertz (60-200 MHz), we have TV frequencies.

We can generalize the above by saying that any conductor of electricity in an alternating electromagnetic field will have an alternating voltage induced in that conductor. Thus, a TV antenna (a conductor) pointed at a TV station which is radiating an alternating electromagnetic field by transmitting a signal will have a signal voltage appear at its terminals. How we utilize and maximize this signal is an important part of the remainder of this book.

The real world is more complex than we have described above. Receiving antennas (except loop antennas) respond to both electric fields and magnetic fields. The electromagnetic field surrounding a transmitting antenna consists of two parts: a near field and a far field. The near field behaves as we described above. The far field, however, does not return to the antenna when current stops, but continues to radiate into space. The far field is the principle electromagnetic field used for transmission and reception of radio waves.

Wavelength

The full range of the different radio frequencies is called a spectrum. It extends from the 10 kilohertz (10,000 cycles per second) to 300 gigahertz (300 billion cycles per second). It is divided into bands of frequencies, as shown in *Table 1-1*. The table also shows how the waves are classified by wavelength. Wavelength (λ) is defined as the distance the wave travels in the time required to complete one cycle. Since, as was stated earlier, the speed is known, the wavelength can be found by the relationship:

$$\text{wavelength } (\lambda \text{ in meters}) = \frac{\text{velocity } (V_1, \text{ meters per sec})}{\text{frequency } (F_1, \text{ cycles per sec})}$$

Note that the higher the frequency, the shorter the wavelength.

Antennas usually are designed to have a length equivalent to one-half, one-quarter or some other sub-multiple (or multiple) of a wavelength. The amount of voltage induced in the receiving antenna depends primarily upon the intensity of the radiated wave, which, in turn, depends mainly upon the transmitting antenna characteristics, and the power at the transmitting antenna.

▶ **Table 1-1 The Radio Frequency Spectrum**

Frequency (in megahertz)	Name of Band	Abbrev	Wavelength (in meters)
0.01 - 0.03	Very low frequency	VLF	300,000 - 10,000
0.03 - 0.3	Low frequency	LF	10,000 - 1,000
0.3 - 3.0	Medium frequency	MF	1,000 - 100
3.0 - 30	High frequency	HF	100 - 10
30 - 300	Very high frequency	VHF	10 - 1
300 - 3,000	Ultra high frequency	UHF	1 - 0.1
3,000 - 30,000	Super high frequency	SHF	0.1 - 0.01
30,000 - 300,000	Extremely high frequency	EHF	0.01 - 0.001

Polarization of a Radio Wave

Figure 1-4 shows the changing electric and magnetic fields produced in an antenna. The term polarization, as applied to antennas, defines the direction of the electric E and magnetic H fields, respectively. The direction of the E field determines an antenna's polarization. An antenna erected so that the radiating element is vertical produces vertical electric waves and, therefore, the antenna is said to be vertically polarized; correspondingly for horizontally polarized antennas. A pure vertical polarized wave will not produce a signal in a horizontally polarized antenna.

Horizontal polarization is standard in the United States for both FM and TV antennas. It was chosen originally because most types of man-made noise are vertically polarized. Thus, horizontal polarization helps reduce interference from such sources.

▶   **Figure 1-4. Magnetic and Electric Fields Around a Current Carrying Conductor**

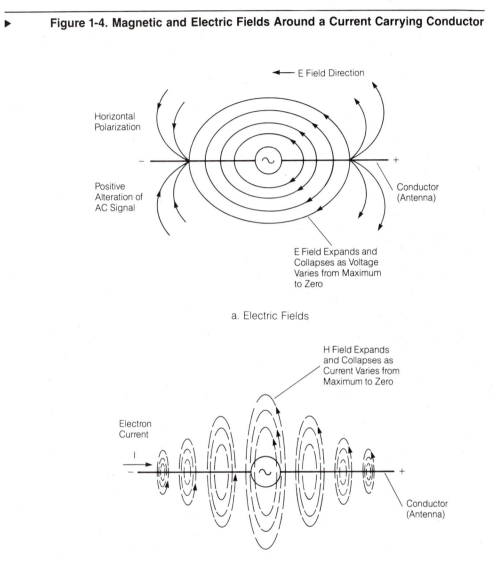

a. Electric Fields

b. Magnetic Fields

Circular Polarization

Because of the increasing popularity of FM portable radios with vertical whip antennas, the FCC has authorized a combination of vertical and horizontal polarization known as circular polarization. In a circularly polarized wave, the E and H fields rotate as they travel through space, making one complete revolution in one wave length (like a corkscrew). Such a wave is received equally well by either a vertically polarized or horizontally polarized antenna.

Recently the FCC has authorized the same combination for TV broadcasting. This results in some improvement in reception by portable television receivers with built-in antennas. Also, a special receiving antenna can be used that is sensitive to the direction of rotation of a circularly polarized wave. This reduces "ghosts" due to reflections because the direction of rotation of circular polarization is reversed when the signal is reflected.

Antenna Impedance

An antenna's impedance, as shown in the following equation, is defined as the ratio of voltage to current at the measuring point:

$$\text{Impedance} = \frac{\text{voltage}}{\text{current}}$$

When the impedance is a minimum, the antenna is said to be resonant at the applied frequency. Under these conditions, for the particular resonant frequency and for a given power, the current in a transmitting antenna is greatest. The impedance of any antenna varies from point to point along its length. This can be seen in *Figure 1-5* which illustrates a plot of current, voltage and impedance along a resonant antenna. The lowest impedance occurs where the current is maximum (at the center).

Figure 1-5. Voltage, Current and Impedance Distribution in an Antenna

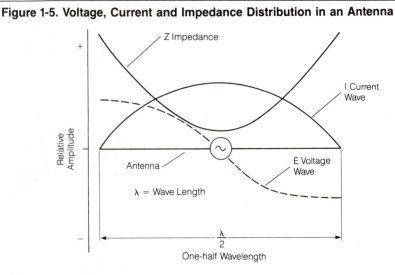

Gain and Directivity

When an electromagnetic wave radiates equally in all directions from a point source in free space, a spherical wavefront results. This is shown in *Figure 1-6a*. Such a source is termed an isotropic source. If a plane is cut through the sphere, as shown in *Figure 1-6b*, the radiation pattern is circular or omnidirectional. An antenna is bidirectional (*Figure 1-6c*) if it concentrates energy in two directions, or unidirectional if it concentrates energy in one direction. An antenna is said to

► **Figure 1-6. Radiation Sources**

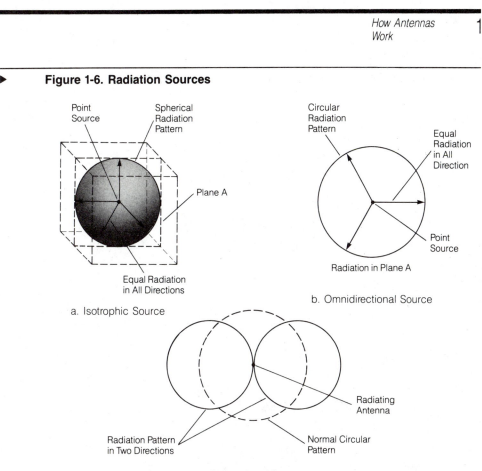

a. Isotrophic Source

b. Omnidirectional Source

c. Bidirectional Source

have gain if the energy radiated in a particular direction is greater than the energy that would be radiated from an isotropic antenna in the same direction. The gain results because of the concentration of energy in the particular direction. Antenna gain is very different from amplifier gain. Power amplifiers output more power than provided to their input. Feeding a power of 100 watts to an antenna does not result in more than 100 watts of radiated energy. Antenna gain is the strength of the radiated field in a particular direction relative to a reference antenna, usually the isotropic source.

Antenna Patterns and Coverage

A radiation pattern is a graph of radiated field strength versus angle or direction from the antenna. The radiation pattern for a horizontal half-wave Hertz antenna is shown in *Figure 1-7*. For this antenna, the pattern shows that maximum field strength occurs at right angles to the antenna while virtually no energy is propagated from the ends. As shown in *Figure 1-7*, the antenna's beamwidth is the angular separation between the half-power points on the radiation pattern.

Another factor that affects the coverage of an antenna is its effective height. This is the height of the antenna's center of radiation above average terrain.

▶   **Figure 1-7. Radiation Pattern for a Horizontal Half-Wave Hertz Antenna**

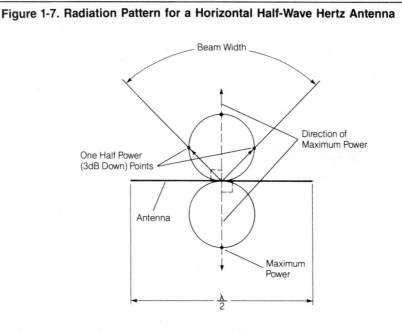

BASIC ANTENNA TYPES

Hertz Antenna

Any antenna complete in itself is known as a Hertz antenna. *Figure 1-7* shows such an antenna. Most FM and TV antennas are Hertz antennas.

Marconi Antenna

A Marconi antenna and its current, voltage and impedance are shown in *Figure 1-8a*. Such an antenna utilizes the earth or other large body (like an automobile) as part of its resonant conductor. The radiation pattern from a Marconi antenna is shown in *Figure 1-8b*. Most low and medium frequency antennas are Marconi antennas. An AM auto radio antenna is a good example.

COMMON TYPES OF TRANSMISSION LINES

There are many types of transmission lines. Various types are shown in *Figure 1-9*. Twin lead and coaxial cable probably are the best known of the types displayed.

Characteristic Impedance

All transmission lines, whether two wire lines or coaxial cables, have a fundamental property called the characteristic impedance. By definition, this is the impedance of an infinitely long piece of the line. The characteristic impedance is uniquely determined by the physical properties of the line — the size of the conductors, their spacing, and the shape and kind of dielectric (insulation) between and surrounding the conductors.

► **Figure 1-8. Marconi Antennas**

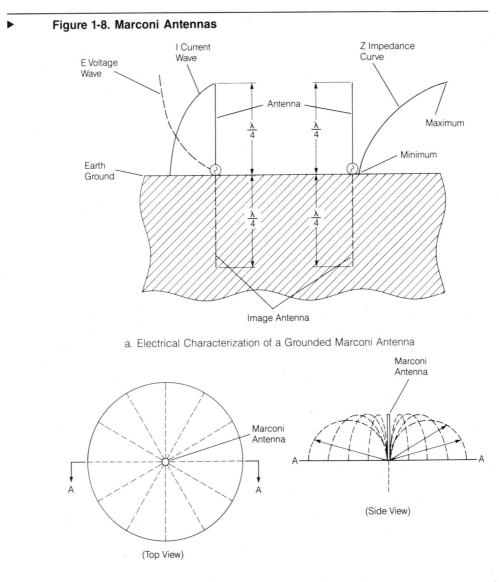

a. Electrical Characterization of a Grounded Marconi Antenna

b. Propagation Pattern for a Grounded Marconi Antenna

Standing Waves

Like reflection of light on a smooth pond of water, when a moving electromagnetic wave encounters a discontinuity in the transmission line, it is reflected. Reflections may be caused by bad connections made through poor connectors, or by a mismatch between the transmission line and antenna, transmitter, or receiver. Reflected power is usually lost, and may produce undesirable side effects which we will discuss later. Reflections mean that energy waves are traveling two ways on the line. These waves interact and produce what are called standing waves of voltage and current on the line. The voltage maximums and minimums of a standing wave have a constant amplitude. The ratio of the maximum to minimum voltage on a line is called the voltage standing wave ratio (VSWR):

$$\text{VSWR} = \frac{V_L \text{ max}}{V_L \text{ min}}$$

If the source and load impedances match the transmission line there are no standing waves and the VSWR, or simply SWR, is 1.

SUMMARY

We now have looked at how antennas work and covered the fundamentals of antennas. Next we will look at TV antennas in general.

▶ **Figure 1-9. Various Types of Transmission Lines**

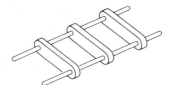

a. Parallel Two-Wire Line

b. Twisted Pair

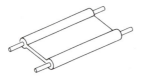

c. Two-Wire Ribbon Flat Lead (Twin Lead)

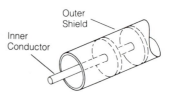

d. Air Coaxial with Washer Insulator

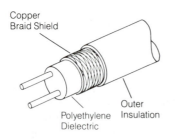

e. Two-Wire Shielded Pair

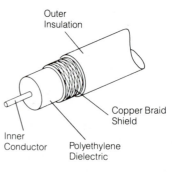

f. Coaxial (Called Coax)

TV Antennas — General

To obtain a sharp, clear TV picture with true color requires the proper TV antenna mounted in the proper place relative to the TV transmitter. In this chapter, we will discuss important factors that contribute to television reception, and the principles of several simple types of TV antennas. The discussion of more elaborate types of antennas continues in Chapter 3. When you finish both of these chapters, you should be well equipped to select a suitable antenna for your particular TV installation.

FACTORS TO CONSIDER IN TELEVISION RECEPTION

TV signals are much more complex than AM or FM radio signals. To understand what factors affect TV signal reception, we must look at the nature of the signal, its frequency allocation, its bandwidth and noise, its signal strength, and its path from transmitter to receiver. Let's look at the nature of the signal first.

The Nature of the TV Signal

We said that the TV signal was more complex than AM or FM radio signals. Look at *Figure 2-1*. It shows the spectrum of signals for a typical TV channel — Channel 3 in this case. Much more information is being transmitted in a TV channel. The picture details, the color information, the sound, and the synchronization signals all are contained in the frequencies from 60 to 66 MHz of Channel 3. Because there is so much more information, TV channel bandwidth is 6 megahertz. AM signals typically require a 10 kilohertz bandwidth, while FM signals require a 200 kilohertz bandwidth.

TV Channel Frequency Allocation

In addition to the greater bandwidth of 6 MHz, the frequency allocations for TV channels cover a very broad range from 54 MHz to 890 MHz. These allocations, shown in *Table 2-1*, are made in three distinct groups of frequencies: low band VHF (54-72 and 76-88 MHz), high band VHF (176-216 MHz), and the UHF band (470-890 MHz). The wide dispersion of frequencies complicates TV antenna

▶   **Figure 2-1. Spectrum of TV Channel Signals**

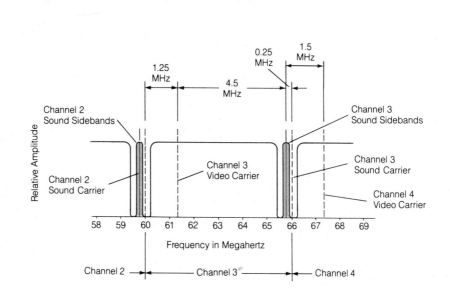

design. It is difficult to design a single antenna that maintains equal performance over the full frequency range. An antenna tuned for the lowest channel frequency is nearly 16 times the size of an antenna tuned for the highest channel!

TV Bandwidth and Noise

The noise in any communication system is proportional to the bandwidth of the information channel. Since the antenna bandwidth is much wider than the TV channel bandwidth, the noise that appears on a TV is limited by the TV channel bandwidth, not the antenna bandwidth. Common examples of noise that is picked up by the antenna and comes through the TV information channel are automobile ignition and commutator brush sparkings.

Signal Strength

We saw in Chapter 1 that a transmitting antenna radiates a magnetic field and an electric field. The signal strength of a radiated signal is expressed in terms of the strength of the electric field. It is a measure of how many volts the electromagnetic field will induce in an antenna that is one meter long. If a receiving antenna were extremely close to a transmitting antenna, the induced voltage would be expressed in volts per meter. However, normally the receiving antenna is a great distance from the transmitter, and the signal strength is only millivolts (thousands of a volt) per meter, or even microvolts (millionths of a volt) per meter.

▶ **Table 2-1. Frequencies of Television Channels**

VHF Band			
Low Band		**High Band**	
Channel	**Freq. (MHz)**	**Channel**	**Freq. (MHz)**
2	54 - 60	7	176 - 180
3	60 - 66	8	180 - 186
4	66 - 72	9	186 - 192
5	76 - 82	10	192 - 198
6	82 - 88	11	198 - 204
		12	204 - 210
		13	210 - 216

UHF Band					
Channel	**Freq. (MHz)**	**Channel**	**Freq. (MHz)**	**Channel**	**Freq. (MHz)**
14	470 - 476	38	614 - 620	62	758 - 764
15	476 - 482	39	620 - 626	63	764 - 770
16	482 - 488	40	626 - 632	64	770 - 776
17	488 - 494	41	632 - 638	65	776 - 782
18	494 - 500	42	638 - 644	66	782 - 788
19	500 - 506	43	644 - 650	67	788 - 794
20	506 - 512	44	650 - 656	68	794 - 800
21	512 - 518	45	656 - 662	69	800 - 806
22	518 - 524	46	662 - 668	70	806 - 812
23	524 - 530	47	668 - 674	71	812 - 818
24	530 - 536	48	674 - 680	72	818 - 824
25	536 - 542	49	680 - 686	73	824 - 830
26	542 - 548	50	686 - 692	74	830 - 836
27	548 - 554	51	692 - 698	75	836 - 842
28	554 - 560	52	698 - 704	76	842 - 848
29	560 - 566	53	704 - 710	77	848 - 854
30	566 - 572	54	710 - 716	78	854 - 860
31	572 - 578	55	716 - 722	79	860 - 866
32	578 - 584	56	722 - 728	80	866 - 872
33	584 - 590	57	728 - 734	81	872 - 878
34	590 - 596	58	734 - 740	82	878 - 884
35	596 - 602	59	740 - 746	83	884 - 890
36	602 - 608	60	746 - 752		
37	608 - 614	61	752 - 758		

The field strength is dependent on the amount of power transmitted, the transmitting antenna gain, the distance from transmitter to receiver, and the path of the signal from the transmitting antenna to the receiving antenna. As we discussed in Chapter 1, this path may be by ground, direct, reflected or refracted sky waves. If the path is short as for a direct wave, the signal field strength will vary very little from day to night. If the path is a reflected wave, the signal field strength can vary significantly between day and night.

Reflections-Ghosts

Signals arriving at the receiving antenna may be other than the direct signal. These indirect signals result when a signal is reflected from tall buildings, water towers, hills, airplanes, etc. *Figure 2-2* shows some examples. It takes only one alternate (indirect) signal path to produce a multiple image on the television screen. This multiple image is called a "ghost" and is illustrated in *Figure 2-3*. The direct signal reaches the receiving antenna first. The indirect signal reaches the receiving antenna at a later time because it follows a longer path. The ghost produced by the indirect signal is the visual equivalent of an echo.

▶ **Figure 2-2. Multiple Reception Paths for a TV Signal**

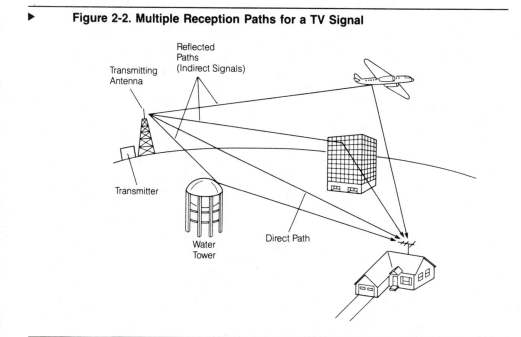

▶ **Figure 2-3. A Reflected (Ghost) Signal on a TV Screen**

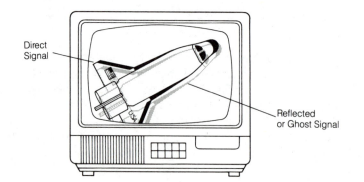

There's also an equivalent echo of the sound signal. However, we do not hear the sound echo because the difference in time between the direct signal and the reflected signal is so small that our hearing cannot detect it. However, on a television screen, the echo effect of the video signal which our eyes see may be very pronounced. We may see a distinct double (or multiple) image which may blur the picture. The directivity of TV antennas can help to reduce TV ghosts significantly. To understand this, let's look at some of the basic types of TV antennas and their characteristics.

BASIC TV ANTENNAS

The Dipole

A basic half-wave dipole antenna is shown in *Figure 2-4*. This simple antenna consists of two metallic rods mounted horizontally. For the frequency to which the antenna is tuned, each rod is a quarter wavelength long, so the total length of the antenna element is a half-wavelength.

One lead-in wire of the transmission line is connected to each rod. The impedance of a simple half-wave dipole at its center is approximately 75 ohms. As discussed in Chapter 1, the impedance of the transmission lead-in also should be 75 ohms to match the antenna so there is maximum energy transfer and minimum reflections.

The basic dipole has limited gain and usually will not give satisfactory results; however, this antenna is the starting point for the more complex types which have better gain and directivity.

► **Figure 2-4. A Simple Dipole Antenna**

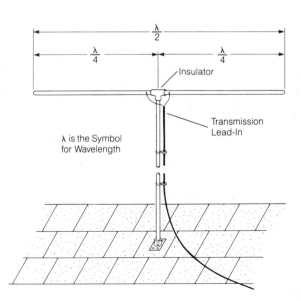

Folded Dipole Antenna

If two half-wave dipoles are placed parallel to each other with their ends connected, the signal currents received in each are in phase. This results in a folded dipole antenna shown in *Figure 2-5*. This antenna has approximately the same gain as the simple dipole, but the frequency response is more uniform over a given band of frequencies. The impedance at the center of the folded dipole is approximately 300 ohms, an increase from the approximately 75 ohms for the simple dipole. "Twin Lead" lead-in 300-ohm transmission line has been designed to match the folded dipole impedance.

Figure 2-5. A Folded Dipole Antenna

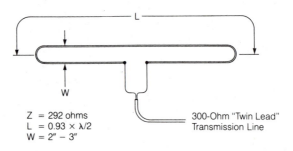

$Z = 292$ ohms
$L = 0.93 \times \lambda/2$
$W = 2" - 3"$

300-Ohm "Twin Lead"
Transmission Line

Yagi Antenna

The Yagi antenna was invented by Professor Uda[1], but was credited to Hidetsugu Yagi[2], a Japanese physicist, because of his English translation of the idea. The Yagi was designed to improve the gain of the antenna concentrated in one direction. The directivity is accomplished with added elements called directors and reflectors. A simple Yagi with a single dipole as the driven element, one reflector, and three directors is shown in *Figure 2-6*. This Yagi has high gain, is very directional, and has a narrow bandwidth. In simple unidirectional antennas like the Yagi, frequency bandwidth is inversely proportional to antenna gain. The higher the gain (and directivity), the narrower the bandwidth.

Antenna Bandwidth

The antenna bandwidth must be adequate to cover all the channels it is designed to receive. This usually means all channels from the lowest VHF to the highest UHF. This cannot be done effectively with a simple antenna. Most TV antennas are a composite of two or more separate antennas integrated into a single structure. The wider the frequency range, and the greater the distance from the transmitter, the more elaborate the antenna structure required. Fortunately, in metropolitan areas, the signal strength is strong enough, so that relatively simple antenna structures are adequate and the more elaborate structures are not needed.

[1] Kennedy, G., *Electronic Communication Systems,* 3rd ed. McGraw-Hill, 1985, p258.
[2] Yagi, Hidetsugu, "Beam Transmission of Ultra-Short Waves," Proceedings of the IRE, Vol. 16, June 1928, p715.

▶ **Figure 2-6. A Yagi Antenna with One Driven Element, Three Directors, and One Reflector**

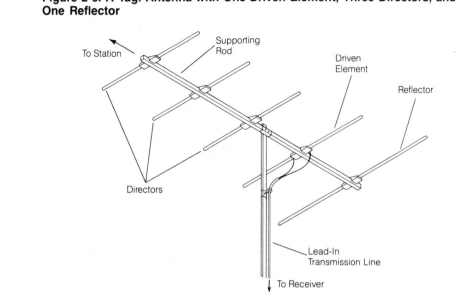

One way to increase the frequency bandwidth of a simple antenna like the Yagi is to increase the diameter of the antenna conductors. The greater the conductor diameter, the wider the antenna bandwidth. Increased conductor diameter also has a second benefit, it increases the physical strength of the antennas. Changing the driven element to a folded dipole also increases the bandwidth. Other antenna designs that have increased bandwidth also can be used. Two of these, the Fan (or Bowtie) antenna and the Conical Array, will be described in Chapter 3.

Directivity and Gain

Directivity is not important for transmitting antennas because they are normally omnidirectional (transmit equally well in all directions). They radiate line-of-sight transmissions, so the transmitting antenna is usually located at a sufficient height to allow for maximum transmission range to the serviced area. Atop the tallest building in a city, or on the highest hilltop in the area, are convenient locations for TV transmitting antennas.

Receiving antennas, on the other hand, are unidirectional because they need to receive a maximum signal from the direction of the TV transmitter. High gain and directivity are closely related. You cannot have one without the other.

Highly directive antennas can be very effective in reducing reception of reflected signals. Let's look at an example. *Figure 2-7* shows a three-element Yagi dipole antenna (driven element, director, reflector) with signals arriving from two opposite directions, A and B. The antenna treats the two signals very differently. Let's consider that direction A is from the TV transmitter and direction B is the source of a reflected signal.

▶ **Figure 2-7. A Three-Element Yagi Receiving Signals from Two Directions**

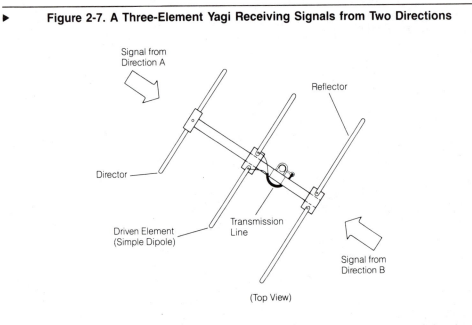

(Top View)

A signal arriving from direction A generates a signal voltage in the driven element, as well as in the reflector and director. Due to the voltage, there is signal current in each of the three elements. The director and reflector re-radiate signals due to the current in them. The element lengths and their distance from the driven element cause the re-radiated signal to induce a signal voltage in the driven element (a simple dipole) that is in phase with the original signal from direction A. This reinforces the signal from direction A, and increases the gain for the signal from direction A.

When a signal approaches from direction B, it also causes additional signals to be re-radiated from the director and reflector. However, the re-radiated signals will arrive at the dipole out of phase with the original direction B signal, producing opposing voltages and reducing the gain from direction B. The relative lengths of the director and reflector and their position cause the respective phase change in the re-radiated signals. As a result, this antenna is much more sensitive to signals coming from direction A, and much less sensitive to signals from direction B.

The net result of the combined actions of the director and reflector is a substantial increase in reception (gain) from direction A, and a corresponding reduction in gain from direction B.

SUMMARY

Now we know something about the nature of the TV signal, its bandwidth, noise, signal strength, reflections and ghosts, and something about the basic antenna and its elements. In Chapter 3, we will add to this knowledge by examining additional antennas and concepts.

Selecting a TV Antenna

In Chapter 1, we described how antennas work. In Chapter 2, we looked at factors that affect TV reception and discussed basic TV antennas. In this chapter we look at more specific TV antenna types, their unique characteristics, and why one would be selected over another.

TYPES OF VHF ANTENNAS AND THEIR CHARACTERISTICS

Basic Forms — Arrays

We have seen that adding reflector and director elements to the basic dipole increases the gain and directivity. These added elements form an array. An array may have as few as two elements, or on some elaborate antennas, as many as fifty elements. We've seen a three-element array (*Figure 2-7*), a five element array (*Figure 2-6*), and *Figure 3-1* is an array of folded dipoles.

Stacked Arrays

Most antennas can be stacked by placing one antenna above another in the same vertical plane. An example is shown in *Figure 3-2*. Stacking provides two advantages: higher gain and increased horizontal directivity.

Theory of Stacking

The antennas that are placed one above the other can be parasitic; however, this is rarely done. The antennas are commonly placed one-half wavelength apart and connected to a common transmission line (lead-in). The spacing between the antennas is not critical, and there are some advantages to much wider spacing — primarily higher gain. The final directional pattern depends on several factors: the number of elements, the spacing between elements, and the phase difference between elements and antennas.

The only time TV antennas are stacked is for the highest gain in very remote locations. Otherwise, the practical difficulties in mounting TV antennas in a stacked array limit their use.

▶ **Figure 3-1. In-Line Antenna of Folded Dipoles**

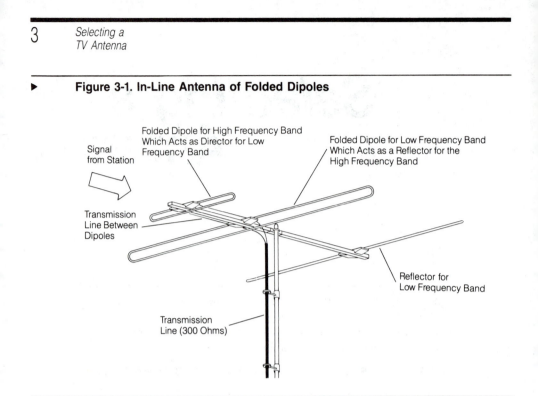

▶ **Figure 3-2. A Stacked Antenna**

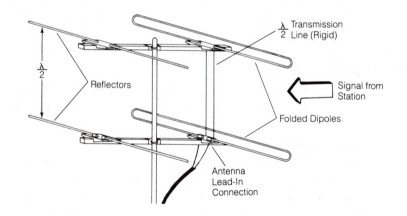

Multi-Element Yagi Antenna

A Yagi antenna contains directors and reflectors. *Figure 2-7* had one of each. In *Figure 3-1*, the dipoles are serving double duty. When not the active driven element, they take on the role of a director or reflector depending on the frequency band. The reflector is normally longer than the driven element, while the directors are shorter. There are several design variations. In *Figure 3-3*, each director is cut and placed as if the previous director were the driven element. This makes the whole structure taper in the direction of maximum gain.

▶ **Figure 3-3. Five-Element Yagi Array with Folded Dipole**

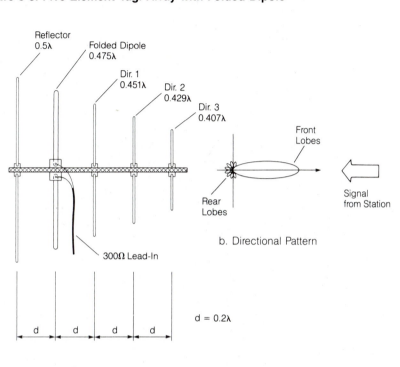

a. Structure

b. Directional Pattern

More than one reflector adds very little to the directivity or gain of the array. However, more directors improve the gain considerably. A three-element Yagi has a directive gain of about 7 dB, whereas a five-element Yagi has a directive gain of about 15 dB. In a properly designed Yagi array, each director adds nearly 1 dB of gain. In the five-element Yagi of *Figure 3-3a*, the driven element is a folded dipole. Its impedance is 300 ohms. Note the highly directional pattern of *Figure 3-3b*.

Log-Periodic Antenna

Antenna research in recent years has led to the development of a new type of broadband, high gain antenna, called log-periodic. This design, shown in *Figure 3-4a*, incorporates a large number of active dipoles connected together in a special phasing arrangement. The active dipoles are spaced in accordance with logarithmic mathematical formula based on the theory of an infinite spiral. The impedance of such an antenna (see *Figure 3-4b*) is a periodic function of the frequency; therefore, the name log-periodic. It has broadband frequency characteristics and its directional characteristics are similar to a dipole array.

► **Figure 3-4. Log-Periodic Dipole Array**

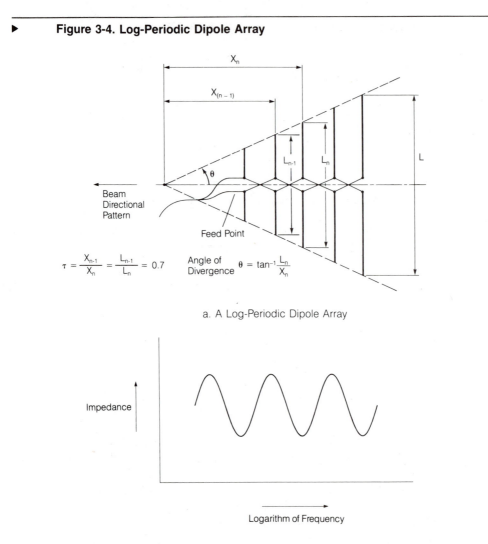

a. A Log-Periodic Dipole Array

b. Periodic Nature of Antenna Impedance on Logarithmic Scale

There is no single style of log-periodic antenna, and the term applies to a variety of antennas having the same electrical characteristics. Two antennas may be very different in appearance, yet both can be log-periodic. In some versions, the elements are bent forward and a UHF section added in front of the VHF section.

Color-Laser Log-Periodic Antenna

The directors of the color-laser antenna are the result of radar antenna design. The directors are circular and resemble a collection of discs on a rod. This gives a uniform gain over a wide frequency band.

Conical Array Antenna

The conical array (*Figure 3-5*) can receive both low band and high band VHF signals. The reflector elements extend straight out from the supporting beam, whereas the front driven elements are bent forward. The response pattern of this antenna contains only one major directional pattern lobe on all channels. A conventional dipole that is sized for a low frequency channel and has a single lobe directional pattern for that channel will have a multilobe pattern for higher frequency channels. In contrast, a dipole cut for high frequency will have poor gain response on low channels. The conical antenna is usable for all the VHF channels.

One way of looking at both a conical array and a UHF fan dipole (*Figure 3-6*) is that both effectively increase the "diameter" of the dipole. This increases the antenna's bandwidth significantly, without the penalty of a clumsy and heavy structure.

TYPES OF UHF ANTENNAS AND THEIR CHARACTERISTICS

The UHF television signals (470-890 MHz) are usually weaker than the comparable VHF signals. Also, the losses in the lead-in cable from the antenna to the receiver are greater for UHF. For these reasons, UHF antennas must be installed more carefully than VHF in order to capture as much signal as possible. Even so, there are many antenna designs to choose from.

▶ **Figure 3-5. Conical Antenna with a Broadband Response Pattern**

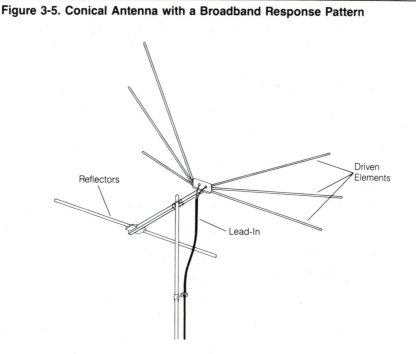

Reflectors

Driven
Elements

Lead-In

UHF Dipole

The most obvious difference between VHF and UHF antennas is the physical size. A half-wave dipole for Channel 2 (55 MHz) will be 10 times longer than one for Channel 28 (550 MHz). This means that a much more elaborate UHF antenna can be constructed without the antenna becoming physically unmanageable. With more elements added to the UHF antenna, higher gain and directivity can be obtained.

Fan Dipole

A UHF fan dipole antenna is shown in *Figure 3-6*. Its directional pattern is basically the same as the common dipole shown in *Figure 1-7*. This antenna is also known as the Bow Tie antenna. By using triangular sheets of metal instead of rods, the bandwidth is greatly increased, covering the entire UHF band. To provide such wide bandwidth characteristics, the fan dipole is slightly longer than the rod dipole.

▶ **Figure 3-6. UHF Fan Dipole Antenna**

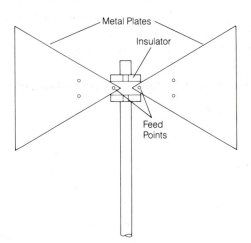

UHF Reflectors

Reflectors have the same action at UHF as at VHF, but increased performance is possible at UHF.

Mesh Reflector

A mesh screen like that shown in *Figure 3-7* is considerably more efficient than a rod reflector. The mesh is preferable because it weighs less and has less wind resistance. The reflector should extend beyond the ends of the dipole, although the dimensions are not as critical as they are for a rod reflector. Mesh screens are as effective as solid metal sheets as long as the mesh spacing is less than about 0.2 wavelength.

▶ **Figure 3-7. Fan Dipole with Mesh Screen Reflector**

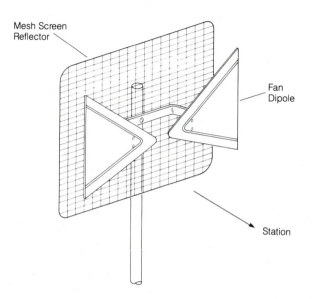

Parabolic Reflector

Reflectors shaped like a parabola are very directive. A parabolic reflector on an antenna concentrates the radiated energy much the same as a parabolic reflector of a flashlight concentrates the light energy. Parabolic reflectors of reasonable size can increase the gain of a simple dipole by as much as 9 dB, which is equivalent to eight times the power. Higher gain is possible, but the size becomes impractical. For sharp vertical directivity, a partial parabolic reflector like the one shown in *Figure 3-8* may be used.

Corner Reflector

The corner reflector is a popular UHF reflector. It has a very high front-to-back pickup ratio; in particular, reception from the backside is greatly reduced. As shown in *Figure 3-9*, the driven element, usually a dipole antenna, is placed at the center of the corner angle. The corner angle affects the power gain, directivity, and impedance. The smaller the corner angle, the lower the impedance. In addition to the corner angle, the distance between the dipole and the vertex of the angle affects the pattern. If the dipole is too close to the corner, the antenna vertical beamwidth will be broadened. If it is too far from the corner, multilobe patterns will result. In *Figure 3-10*, the corner reflector is combined with Yagi type directors to increase gain and directivity of UHF TV antennas. The gain of the corner reflector antenna is high over the entire UHF band. Typical antennas range from about 7 dB (5 times) at the low end to 13 dB (20 times) at the high end.

► **Figure 3-8. A UHF Antenna Consisting of a Dipole and a Partial Parabolic Reflector**

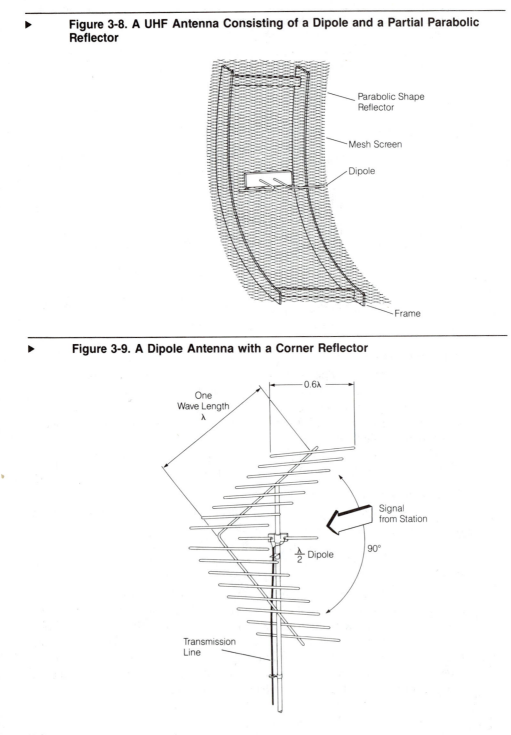

► **Figure 3-9. A Dipole Antenna with a Corner Reflector**

▶ **Figure 3-10. Corner Reflector Antenna for UHF TV** *(Courtesy of Radio Shack)*

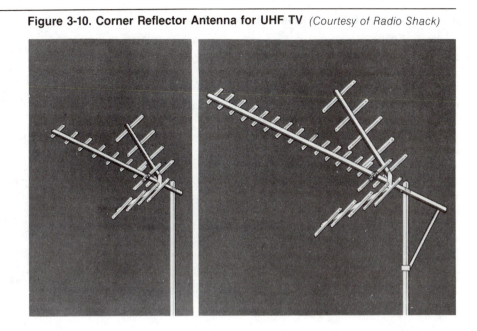

COMBINATION ANTENNAS

Because of the increased popularity of UHF TV channels, many areas now have both VHF and UHF TV stations. In these areas, combination antennas make the installation task much simpler.

Most electronics stores offer a wide variety of powerful "three-in-one" (VHF, UHF, FM) antennas for all reception areas. These advanced-design combination antennas employ a variety of features and types such as the corner reflector, conical swept elements, multi-array Yagi, and the log-periodic design.

Advantages of Combination Antennas

Combination antennas are very popular for either new or replacement antenna installations. These antennas feature high gain, good directivity and broad bandwidth. Some antennas with front to back ratios between 30 and 40 dB are capable of providing clear reception and sharp pictures even in weak signal locations or areas of high noise interference. These antennas are mechanically rugged and can withstand strong winds and icing conditions.

Figure 3-11 shows a combination antenna for VHF, UHF and FM. A three-way splitter (*Figure 3-12*) is used to separate the VHF, UHF and FM signals so that they can be fed to their appropriate receivers. Only one down lead-in is required.

► **Figure 3-11. Combination FM and TV Antennas** *(Courtesy of Radio Shack)*

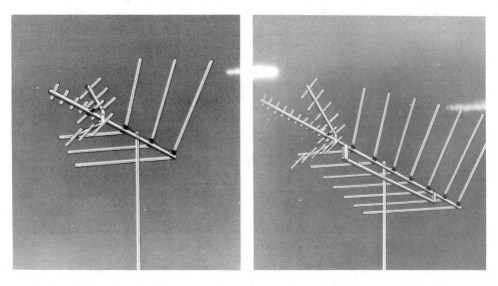

► **Figure 3-12. Signal Splitter** *(Courtesy of Radio Shack)*

INDOOR ANTENNAS

VHF Antennas

Indoor VHF antennas are for use in fairly strong signal areas. A simple V-type model with telescoping rods that extend to 35 inches (called "rabbit ears") sells for under $5.00 *(Figure 3-13a)*. A more elaborate model *(Figure 3-13b)* has a 12-position switch which can modify the response pattern, raise and lower the resonant frequency of the antenna, and minimize ghost signals.

High Efficiency Models

A highly efficient model *(Figure 3-13c)* is available with an added preamplifier which provides up to 20 dB amplification for high quality reception. Each telescoping rod has an axial inductor near its lower end to improve performance on VHF and FM. The unit has a switch to turn the amplifier on and off, a gain control, and a fine tuning control.

▶ **Figure 3-13. Indoor TV Antennas** *(Courtesy of Radio Shack)*

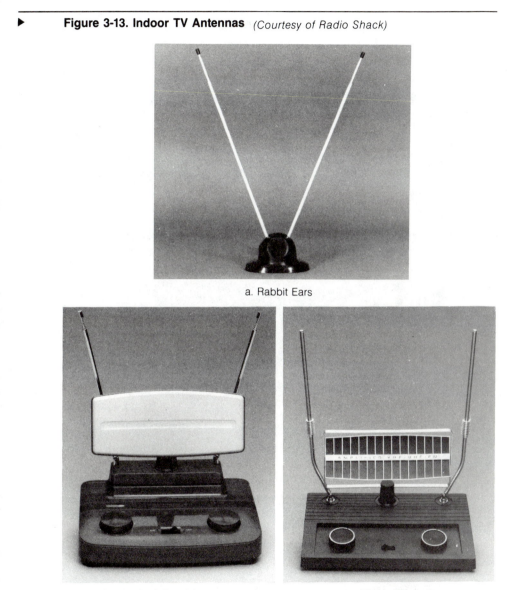

a. Rabbit Ears

b. Adjustable   c. Highly Efficient

UHF Antennas

Indoor UHF antennas also are satisfactory only in strong signal areas. These antennas vary from a simple loop about 7 inches in diameter on the back of the TV, to a fairly elaborate configuration of various size loops, with a means of adjusting the phase shift between the loops.

Another indoor UHF antenna that is particularly efficient is an upright twin bow-tie antenna. One type is shown in *Figure 3-14.*

▶ **Figure 3-14. Upright UHF Twin Bow-Tie Antenna** *(Courtesy of Radio Shack)*

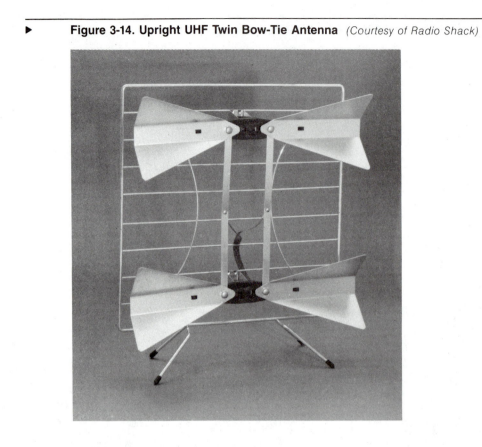

SUMMARY

Some of the antennas described in this chapter are capable of receiving TV signals from over 200 miles away. In the next chapter, we'll look at receiving television from over 22,000 miles away as we consider satellite television dishes.

Satellite Television Antennas

The U.S. satellite television system was established originally for cable TV (CATV) systems. It was not envisioned for home TV reception. Now, Home Satellite TV (HSTV) systems are becoming commonplace. Over 100 news, sports, entertainment and special services programs may be received from approximately 20 satellites 24 hours a day. Future predictions indicate that most homes will have their own satellite TV receiving system. Such HSTV systems also are referred to as television receive only (TVRO) systems because they only receive information; they do not transmit it.

SATELLITE COMMUNICATIONS

A communication satellite system like the one shown in *Figure 4-1* is essentially a microwave link repeater. It consists of a satellite, a ground-based transmitting station (uplink) and an earth station or stations for reception (downlinks). The satellite itself is composed of antennas, receivers and transmitters that perform the communications signal transponding (transmitting-responding) function. Satellites may have up to 24 transponders or relay systems which are similar to channels on a TV set. Satellites are powered by solar cells which also keep batteries recharged so the satellite can operate when the earth shades it from the sun.

Orbits

The positions of the Domestic Communications Satellites (DOMSATS) are shown in *Figure 4-2*. They are 22,300 miles above the earth's equator in what is referred to as the "Clarke Belt," and are traveling at 7,000 miles per hour (2 miles per second). This gives them the same angular velocity as the earth; that is, one revolution every 24 hours. As a result they are said to be geostationary. They hover at a fixed point above the equator. They are in what is called a geo-synchronous orbit. Receiving antennas can be aimed at a fixed point in the sky.

▶ **Figure 4-1. Satellite Communication System**

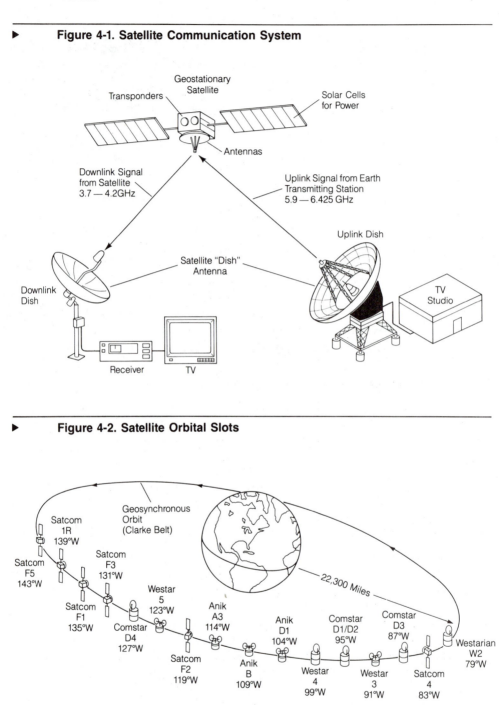

▶ **Figure 4-2. Satellite Orbital Slots**

Uplink and Downlink Frequencies

Satellite programming originates in a studio and is carried over transmission lines to a large transmitting antenna. This antenna is aimed precisely at the position of the desired satellite in the geosynchronous orbit. Most TV satellites operate in the microwave C band at this time. The earth-to-satellite transmission (uplink) frequencies are 5,900 MHz to 6,425 MHz. The satellite-to-earth (downlink) frequencies are from 3,700 to 4,200 MHz. The television distribution bands are shown in *Table 4-1*. Extensive use of the Ku band (12-14 GHz) is proposed. This will undoubtedly accelerate the use of HSTV systems because the antenna size would reduce from 6 to 10 feet in diameter to 2 to 3 feet in diameter. Because of the higher frequencies, components for the Ku band are more difficult to manufacture, and thus, more expensive.

▶ **Table 4-1. TVD (Television Distribution) Frequency Bands**

P band or UHF	200 to 400 MHz	X band	7250 to 7750 MHz
L band	1530 to 2700 MHz		7900 to 8400 MHz
S band	2500 to 2700 MHz	Ku band	10.95 to 14.5 GHz
C band	3400 to 4200 MHz	Kc band	17.7 to 21.2 GHz
	4400 to 4700 MHz	K band	27.5 to 32.0 GHz

COMPLETE HOME SATELLITE TV SYSTEM

The basic components of a home satellite TV system, some of which are shown in *Figure 4-3*, include:
1. an antenna dish;
2. a feed horn;
3. a low-noise amplifier/block converter (LNA, LNB, LNC);
4. the HSTV receiver;
5. antenna positioner;
6. stereo processor and audio system; and
7. accessories such as VCRs and remote controls.

Antenna

The antenna (*Figure 4-3a*) is a radio reflective, shallow parabolic antenna that concentrates received signals to a focused point. It can be made of fine mesh wire, or be all metal, or have a shell of fiberglass with metal or wire mesh imbedded in it.

Feed Horn

Figure 4-3b shows a feed horn. They are made in various shapes—square, rectangular or round. The feed horn is a very important part of the receiving station. It is mounted at the antenna focal point and captures the signal from the reflective dish.

► **Figure 4-3. Home Satellite TV System**

a. Parabolic Antenna Dish
(Courtesy of Channel Master®)

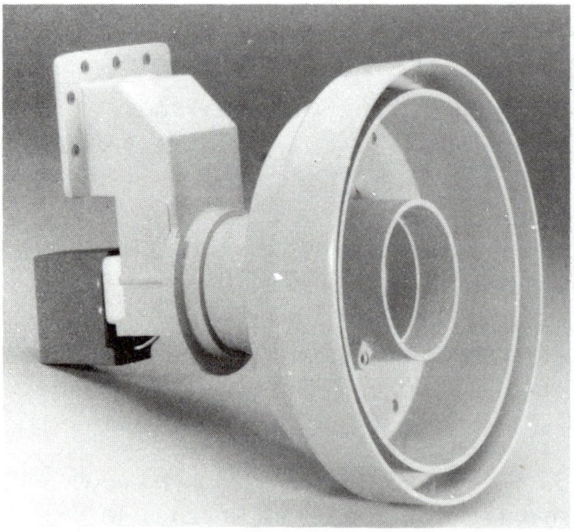

b. Feed Horn
(Courtesy of Radio Shack)

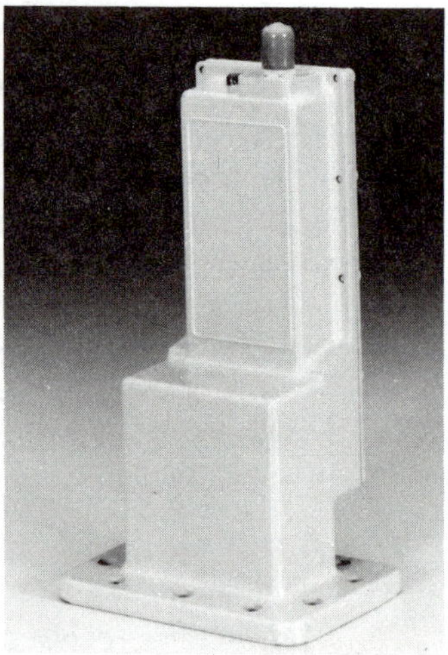

c. Low-Noise Amplifier-Block Converter
(Courtesy of Radio Shack)

d. **HSTV Receiver, Positioner, and Descrambler**
(Courtesy of Channel Master®)

Transponders in the satellite retransmit their signals with either vertical or horizontal polarization. Feed horns must correspond to receive the signals. Two methods are used to get correspondence. The most expensive method is to use a feed horn designed to couple to a dual LNA/LNB—one for the vertical signals and one for the horizontal signals. A much less expensive method is to simply turn the feed horn assembly 90°, which can be done with a standard, low cost TV antenna rotor.

Low-Noise Amplifier and Converter

The LNA/LNB/LNC (*Figure 4-3c*) is mounted to the feed horn. It receives the signals from the antenna and amplifies them up to 100,000 times. One can purchase low noise amplifiers and down converters separately, but it is standard practice for the down converter to be contained in the LNA (called an LNB or LNC) to avoid transmission line loss.

Receiver

Figure 4-3d shows the satellite receiver, the system component inside your home which allows you to select channels from the satellite. The receiver has features such as on-screen selection display, programmed digital stereo on VCII channels, as well as parental channel control.

MICROWAVE ANTENNAS

Design Theory Considerations

The antennas used for microwave signals bear little resemblance to the antennas we have considered so far. Because of the short wavelength involved, the physical sizes required are small enough to allow quite different arrangements that are not practical at the lower TV frequencies. Microwave communications are generally from point-to-point rather than the omnidirectional pattern of broadcast stations.

Generally, the power levels that are used at microwave frequencies are limited because of the availability and high cost of microwave power devices. This, as well as the increased noise of the devices at microwave frequencies, is compensated for by highly directional antenna systems, both for uplink and downlink legs.

Parabolic Reflector Dish

We have all seen light beams focused in flashlights and automobile headlights. The reflector is a parabolic reflector. The same principle is applicable to microwave energy. The only restriction is that the diameter of the paraboloid be at least 10 wavelengths at the lowest operating frequency. The basic principles and specifications of a parabolic dish antenna are shown in *Figure 4-4*. The feed horn focal length results from choosing the diameter and the depth. The diameter is set by the range of frequencies that the antenna must receive. Different ways of coupling signals to and from the antenna dish are shown in *Figure 4-5*. The

Cassegrain feed (*Figure 4-5c*) is used to shorten the feed mechanism length in critical applications. It uses a hyperboloid secondary reflector that has a focus point coincident with that of the paraboloid.

▶ **Figure 4-4. Parabolic Dish Antenna**

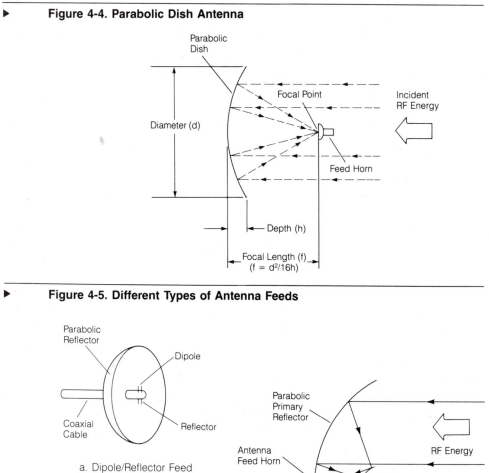

▶ **Figure 4-5. Different Types of Antenna Feeds**

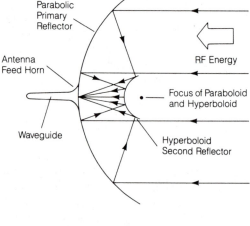

Each type of dish—solid, mesh, metallic, fiberglass—has its own benefits. Wind resistance, cost, ease of installation, quality, etc., are just a few of them. An aluminum dish will maintain its parabolic shape and efficiency for years. The sun and weather can have very pronounced effects on fiberglass which causes cracking, separation of layers, loss of shape and efficiency.

Mountings

As we discussed previously, the twenty or so satellites are in stable positions along the equatorial Clarke belt at an altitude of 22,300 miles. In order to receive a particular satellite, the receiving dish must be pointed precisely at the satellite.

Azimuth-Elevation

Two types of mounts are in general use. The Azimuth-Elevation (AZ-EL) mount is shown in *Figure 4-6a*. A movable dish with an AZ-EL mount allows rotation in the horizontal plane from due north to set the azimuth angle (θ_{AZ}). A separate adjustment provides the correct elevation angle (θ_{EL}).

Polar

Only one adjustment is needed using the polar mount shown in *Figure 4-6b*. The hour axis is positioned to sweep through the arc of the synchronous orbit of the satellites. One adjustment allows positioning to aim at a particular satellite. When the antenna was mounted originally, the elevation angle and main support for the antenna were adjusted permanently to point at the North Star. Advanced

▶ **Figure 4-6. Satellite Receiving Antenna Mounts**

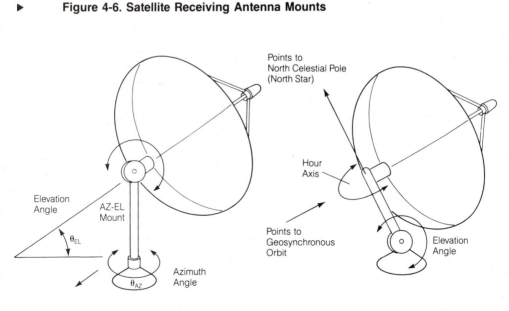

a. Azimuth-Elevation (AZ-EL) Mount   b. Polar Mount

design motorized mounts are computer-controlled from the receiver. You simply select the desired channel with the wireless remote control and the receiver's built-in computer does the rest and positions the dish precisely. The receiver shown in *Figure 4-3d* has computer control capability.

Low-Noise Amplifiers and Converters

LNA stands for low-noise amplifier. Signal strengths are very low and transmission line loss is very high unless expensive waveguides are used. Therefore, the LNA is located as close to the focal pickup point of the antenna as possible. It receives the very low-level signal from the feed horn and amplifies it to a satisfactory level for the receiver. Also, unfortunately, it amplifies its own noise. Low-noise amplifiers are rated in degrees Kelvin. The lower the noise, the lower the rating, the better the amplifier—perfect being 0° Kelvin. A good LNA is one in the 30° to 60° range.

After the LNA, the received signal frequency (in gigahertz) must be converted down to the final intermediate frequency (usually 70 MHz). A low-noise converter (LNC) or a low-noise block converter (LNB) provides this function. As stated previously, in most cases the low-noise converter is located near or is an integral part of the LNA. The output of a converter can be coupled through a smaller and less expensive coax cable to the receiver than the expensive coaxial cable required to use with an LNA. Most LNC also provide additional amplification besides down conversion to a lower frequency.

A low-noise block (LNB) converter is not limited to a single channel, but converts the entire group of frequencies of one polarization received by the TVRO or HSTV to a lower frequency group accepted by the receiver. The entire band is converted at once, not just isolated frequencies as in an LNC.

SUMMARY

The satellite TV industry is still young but expanding rapidly. Technology is changing every day to bring the world closer together. The time will come when most homes will have their own HSTV system and be a part of a continentwide or even worldwide communication network. The material covered in this chapter should give you a good foundation in the antenna systems used to make that possible. In the next chapter, we'll discuss antennas used for FM reception.

FM Antennas

BACKGROUND

In 1933, Edwin H. Armstrong, the inventor of the superheterodyne circuit used in virtually all radio, radar, and TV receivers, invented frequency modulation (FM). In frequency modulation, the sound input signal varies the transmitted carrier frequency rather than the amplitude. As a result, FM reception will be virtually noise-free if there is adequate signal strength in the area.

A second advantage of FM over AM is that the fidelity of an FM broadcast is superior. High fidelity signals and FM require greater channel bandwidths to handle the signal information. Noise increases with greater bandwidth; therefore, FM channels should have increased noise. However, proper system design keeps the noise low. But there is a price that must be paid. FM channel bandwidth is 200 KHz, while AM channel bandwidth is 10KHz. Twenty AM stations will fit in the same frequency spectrum as one FM station. The consequence is that FM carriers must be much higher in frequency than AM carriers in order to carry the number of channels and the wide bandwidth information.

The FM Band — Its Frequencies and Wavelengths

In this chapter we will consider antennas that receive FM signals that are used in these five major service areas:

1. FM radio
2. Television audio (sound)
3. Satellite TV (both video and audio)
4. Public Service Radio (police, fire, taxicabs, etc.)
5. Amateur radio.

FM broadcast radio in North America is allotted frequencies in the VHF band between 88 and 108 MHz. This is between the low and high TV VHF bands (Channel 6 and 7 *Table 2-1*). Within this band, the stations are spaced 200 KHz apart. A full wavelength at the end frequencies is given in *Table 5-1*.

▶ **Table 5-1. FM Wavelengths**

Frequency (megahertz)	Wavelength (meters)	Wavelength (inches)
88	3.409	134.2
108	2.777	109.4

$$\text{Wavelength in meters} = \frac{300 \times 10^6 \text{ meters per second}}{\text{frequency in cycles per second}}$$

From *Table 5-1*, it is apparent that half-wavelength dipoles are 1.3 meters to 1.7 meters (55 to 67 inches) long.

TYPES OF FM ANTENNAS

Simple Dipole

A simple dipole antenna may be used for FM reception in strong signal areas. A dipole antenna, whose length is a half wavelength of a mid-band frequency, will have a reactive component which will cause a slight mismatch into a 75 ohm load. Shortening the dipole about 5% reduces the reactive component to zero, and provides a closer match into a receiver's 75-ohm antenna input terminals.

Folded Dipoles

The folded dipole antenna has been a favorite FM antenna for years. A very simple, inexpensive antenna for indoor use in high signal strength areas is shown in *Figure 5-1*. It can be connected easily to the receiver and mounted conveniently nearby. You can make a similar one from 300-ohm twin lead as shown in *Figure 5-2*.

▶ **Figure 5-1. FM Dipole Antenna**

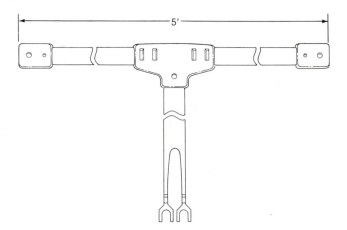

► **Figure 5-2. Simple Homemade Indoor FM Folded Dipole Antenna**

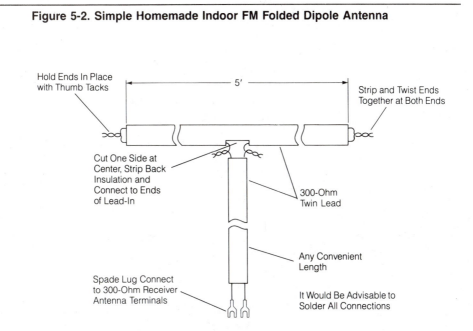

Figure 5-3 shows another antenna that can be made at low cost. It is a folded dipole that has good performance characteristics. By using ¼-inch to ⅜-inch diameter tubing, the antenna will be self-supporting and rigid. It will have a bandwidth sufficient to cover the total FM broadcast band (88-108 MHz). All of these antennas have terminal impedances of approximately 300 ohms, which matches standard 300-ohm twin-lead transmission line. Modern receivers have both 300-ohm and 75-ohm inputs.

Folded dipole antennas can be used for omnidirectional reception. This is illustrated in *Figure 5-4*. In *Figure 5-4*, two dipoles are placed at right angles to each other to produce a near circular reception pattern. In *Figure 5-5*, a single folded dipole with an amplifier boosts UHF and VHF signals by as much as 20 dB to pull in signals up to 90 miles away.

Yagi-Uda

The Yagi antenna for TV was described in Chapters 2 and 3. *Figure 5-6* shows a version of the Yagi antenna that is designed especially for fringe area reception in the FM broadcast band. This antenna differs from the standard TV Yagi in that it has three driven elements rather than one. Such an antenna has very high gain and directivity, so it is capable of receiving FM signals from stations over 100 miles away.

▶   **Figure 5-3. Low Cost "Do-It-Yourself" Folded Dipole**

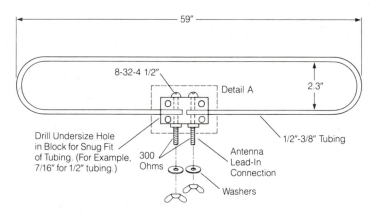

a. Side View

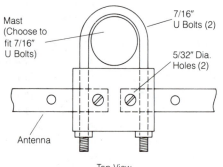

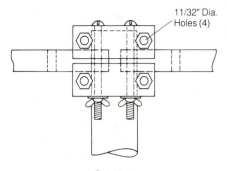

b. Detail A

▶ **Figure 5-4. Omnidirectional Folded Dipoles for FM (Mounted at Right Angles)**
(Courtesy of Radio Shack)

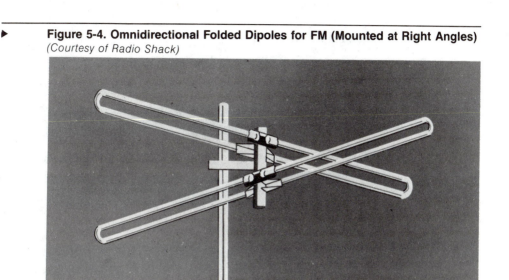

▶ **Figure 5-5. Folded Dipole with Amplifier** *(Courtesy of Radio Shack)*

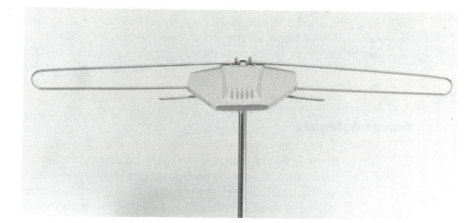

► **Figure 5-6. Yagi FM Antenna** *(Courtesy of Radio Shack)*

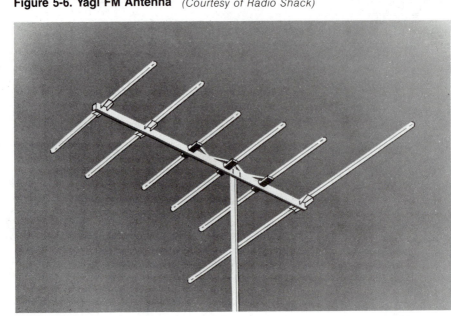

Combination Antennas

Since the FM broadcast band (88-108 MHz) is between the VHF TV low and high bands, broadcast TV antennas are usually designed to include this group of frequencies. For example, the TV antenna shown in *Figure 3-11* features wide-swept elements to capture the maximum amount of signal from TV channels 2 through 83. Such a combination antenna with a suitable splitter gives excellent FM/AM stereo reception, even picks up those "hard-to-get" stations.

Indoor Antennas

FM receivers sometimes come with a built-in antenna consisting of a piece of wire stapled to the inside of the cabinet to form an approximation of a folded dipole. If the signal strength is strong enough and the receiver is properly located and positioned, the resulting performance may be acceptable. For better performance of FM stereo without drift and noise, even in a strong signal area, an indoor antenna like the one shown in *Figure 5-7* may be advisable. The fine tuning knob improves station selectivity and channel separation, and will enhance FM stereo functional effectiveness.

Passive Splitters

If you chose to use a combination antenna for both TV and FM, all signals come from the antenna on one lead-in. The signals must be split to separate them for the receivers. Typical splitters are shown in *Figure 5-8* and *Figure 5-9*. They are passive splitters and have no amplification. To obtain good results, the received signal must be fairly strong. The splitters separate the composite signal into its UHF, VHF, and FM parts. They are inexpensive and convenient to use, and are available in a variety of combinations of 300 and 75-ohm impedances. *Figure 5-8* has an input impedance of 300 ohms and splits the signal to 300 ohms and 75 ohms. *Figure 5-9* has the same output, but has an input impedance of 75 ohms.

▶ **Figure 5-8. Passive Splitter (300 Ohm to 300 Ohm and 75 Ohm)**

(Courtesy of Radio Shack)

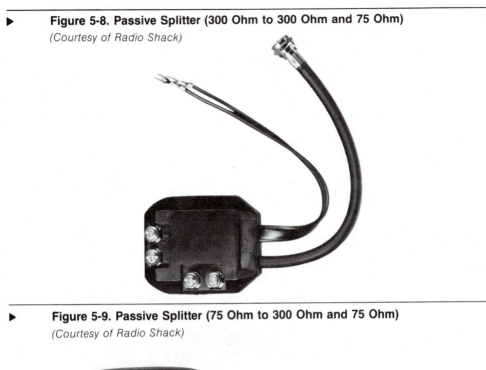

▶ **Figure 5-9. Passive Splitter (75 Ohm to 300 Ohm and 75 Ohm)**

(Courtesy of Radio Shack)

SUMMARY

In this chapter we have discussed antennas especially for FM, completing the discussion of antennas for TV and FM, particularly in strong signal areas. In the next chapter, we will deal with antenna installations in weak signal areas.

Fringe Area and MATV Antennas

Chapter 2 discussed factors that affect TV reception and basic TV antenna types. Chapter 3 and 5 reviewed more specific TV and FM antennas that will be located in areas with satisfactory signal strength. In this chapter, we will consider the problems of antennas located in weak signal areas — the so-called "fringe areas." In fringe areas, you might deal with noise (snow), interference, flutter, and ghosts. You must find ways to reduce or eliminate these effects by using high gain antennas, booster amplifiers, and antenna rotors.

Also, we will see how a master antenna system (MATV) is used to supply a TV signal to multiple sets. These systems originated in fringe areas. They provided good signal distribution from the highest point in a community to TV sets throughout the community. But now the techniques have been applied to local commercial installations, and even to better signal distribution in the home.

FRINGE AREA RECEPTION

The area just beyond the limits for reliable signal reception from an FM or television station transmitter is referred to as the fringe area.

The dividing line between adequate signal strength areas and fringe areas is normally a contour line beyond which the signal strength is less than 500 microvolts per meter. The location and shape of this contour line depends on such factors as:

1. the power of the transmitter;
2. the height of the transmitting antenna;
3. the radiation pattern of the transmitting antenna;
4. the topography of the land (shielding by hills and valleys); and
5. the shielding effect and reflections from water towers, power lines, buildings, weather in the transmission path, etc.

All of the above affect the received signal strength. Since line of sight transmission is so important to TV and FM reception, usually the higher the receiving antenna is in the fringe area, the better.

Problems in the Fringe Area

In the fringe area of a station, the signals are weak, and random noise in the signal may cause snow in the picture. Signals are often erratic causing the picture to be unstable. In extreme fringe areas, the picture may fade out completely, either momentarily or for extended periods. Multiple images (ghosts) are common because signals arrive by multiple paths (see *Figure 2-2*) after reflection from buildings or other obstructions. Variable reflections from moving objects, particularly aircraft, may cause a flutter in the picture. Movement of weather fronts can wipe out the picture altogether.

TV transmitting antennas serving metropolitan areas are usually grouped close together. As a result, a receiving antenna can be pointed in one direction and receive all channels. In fringe areas, this is not necessarily true because a receiving antenna maybe located in the fringe area of several transmitters; therefore, reception differs as the antenna direction is changed. Either a set of multiple antennas, or some convenient method for pointing a single antenna, is required to select the strongest signal.

The quality and uniformity of reception in fringe areas may be greatly improved by using high gain, highly directional antennas, possibly with an antenna mounted preamplifier, and high sensitivity, low noise receivers.

The TV Receiver

A properly functioning TV receiver will give a satisfactory color picture with as little as 50 microvolts of signal at the antenna terminals. For smaller signals, the receiver has to be especially sensitive to give acceptable results. If the signal strength drops below a noise threshold — the level of ambient random (atmospheric) noise that causes "snow" in the TV picture — further increases in receiver sensitivity will not help. Both the noise and the weak TV signal are amplified together. More amplification results in more noise at the same time.

FRINGE AREA ANTENNA SYSTEMS

The solution for weak signals is to use the best antenna system possible — the antenna that will produce the strongest signal on each of the desired channels. This results in the highest signal-to-noise ratio. How is this accomplished? In many cases, several solutions are possible. In particular cases, a combination of more than one solution is necessary.

High-Gain Antennas

In previous chapters, we showed that the gain of an antenna increases with the number of elements it possesses. Therefore, fringe antennas have a large number of elements. The gain of these multi-element antennas may be as high as 15 or 20 dB. (Each 3 dB increase of antenna gain doubles the power of the received signal). Since such a high gain antenna is very directional, if it is not aimed properly, the received signals will be much smaller than they should be. This is despite the fact that the antenna has high gain.

Figure 6-1 shows an antenna suitable for fringe-area reception. This antenna has a boom length of 160 inches. The high gain and directionality is a result of its 48 elements. It's VHF elements are placed at an angle of 60 degrees in what is called a "swept-design." The UHF Yagi has a corner reflector which increases its sensitivity even more. This antenna, if properly installed, should bring in UHF stations over 100 miles away and VHF stations from nearly 200 miles away.

► **Figure 6-1. A High Gain, Directional, Wide Bandwidth Antenna**

(Courtesy of Radio Shack)

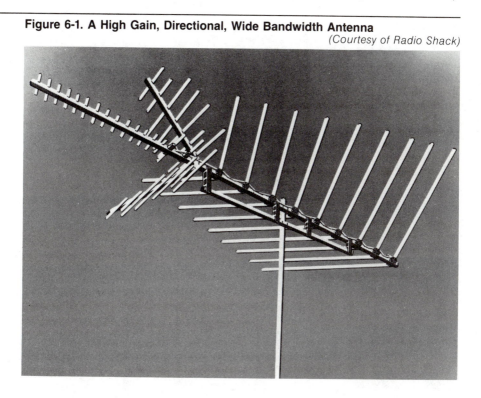

Multiple Antenna Installations

In fringe areas all of the channels will be weak. However, one channel may be weaker than the others, and an antenna with maximum gain for that channel frequency may be selected. If necessary, several such antennas can be stacked and the entire array raised as high as possible above any other part of the antenna system. An example is shown in *Figure 6-2*. Usually the higher the antenna, the stronger the received signal. However, there are heights at which signal cancellation occurs because of waves reflected from the ground. It is a good idea to try different heights for the antenna to get the best reception, especially in hilly areas.

► **Figure 6-2. Two Double-Stacked Arrays Separated on a Tower**

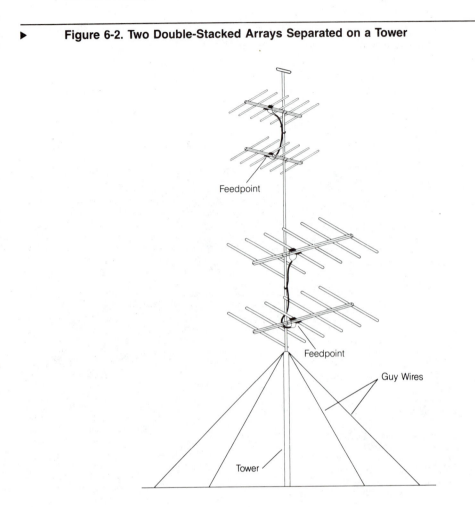

Antenna Preamplifiers

If the antenna is mounted on a very tall tower, the transmission line may be quite long — possibly several hundred feet. Transmission line is not perfect, and the losses in a long line will reduce the signal significantly at the set. In this case, an outdoor preamplifier (*Figure 6-3*), mounted on the antenna mast as close to the antenna as possible, will improve performance substantially. This preamplifier is a broadband transistor RF amplifier. The DC power for the amplifier is supplied by an AC power line converter, and is fed through the same transmission line that sends the signal to the TV set. The preamplifier boosts the TV signal from the antenna and improves the signal strength and signal-to-noise ratio at the input to the TV set.

► **Figure 6-3. A Mast-Mount TV/FM Antenna Signal Amplifier**
(Courtesy of Radio Shack)

Antenna Rotors

When TV broadcast stations are located in different directions from the receiving
set, an antenna rotor provides a convenient means for pointing the receiving
antenna at the different channels. Multiple path reception problems (i.e. ghosts)
also may be reduced or eliminated by using an antenna rotor.

Figure 6-4a shows a rotor motor mounted on a mast which is located on a
roof. The control module (*Figure 6-4b*), located near the TV receiver, furnishes
power to the rotor by means of a cable. Using the control module, the rotor is
activated, and while watching the set, the antenna is rotated for the best
reception.

Most control modules indicate the direction in which the antenna is pointing. This allows the best direction for each channel to be logged, and the
antenna quickly returned to the respective channel direction the next time a
particular channel is selected. Most rotors have brakes which stop rotation
quickly, and also will hold the antenna stationary in high winds.

Antenna rotation is slow enough so that the best direction for a particular
station is readily determined. Rotation stops are provided to prevent wrapping
the cable around the mast. Some control modules have lights that indicate when
the antenna has been rotated to one of the rotation stops.

▶ **Figure 6-4. A Mast-Mounted Rotor and Its Control Module**

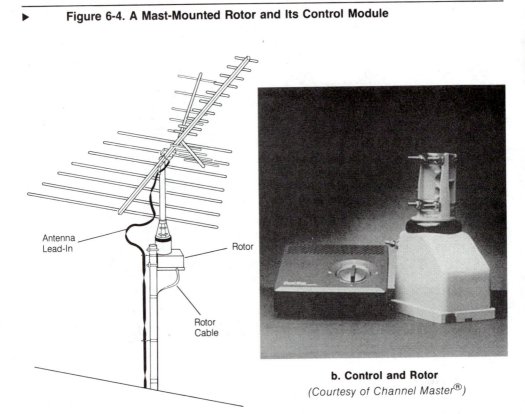

b. Control and Rotor
(Courtesy of Channel Master®)

a. Mounted Rotor

MULTIPLE SET DISTRIBUTION

Experience gained from Master Antenna Television systems (MATV) in fringe areas has been applied to commercial and home systems in strong signal areas. Even in a home, you may wish to connect two or more TV sets to a single antenna. Unless properly connected, picture quality can be degraded because of power loss from impedance mismatch at the interconnection points, and the resulting reflections that occur. These problems can be eliminated by using suitable couplers (with preamplifiers or booster amplifiers, if needed).

Simple Home Systems

In strong signal areas, a single antenna will provide signal to two or more TV sets using simple passive couplers such as the ones shown in *Figure 6-5a* and *6-5b*. They come in several varieties. One antenna can feed signals to 2 or 4 sets. Some can be used both indoors and outdoors. Different units can be selected to properly impedance match 300-ohm twin lead or 75-ohm coax cable.

► **Figure 6-5. Passive Couplers** *(Courtesy of Radio Shack)*

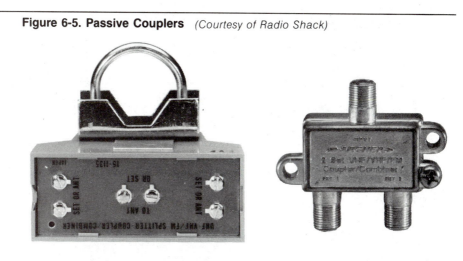

a. Indoor/Outdoor, 2 Set, 300 Ohm   b. Indoor-Outdoor, 2 Set, 75 Ohm

In moderate or low signal areas, a preamplifier or booster amplifier normally is required to prevent reduced picture quality. There are two basic types of booster amplifiers: (1) a mast mounted antenna preamplifier (*Figure 6-3*) that is mounted as close as possible to the antenna, and (2) a booster amplifier normally mounted at a convenient location inside the home. The booster amplifier may include a two or four set coupler.

Booster Amplifiers

In effect, a booster amplifier is like adding a stage or more of RF amplification to the TV receiver, but placed ahead of the distribution point. Because it strengthens the signal, it may increase the contrast range of the picture and improve the signal-to-noise ratio of a poorly performing set.

A block diagram of a multiple receiver system is shown in *Figure 6-6*. This uses a booster amplifier (*Figure 6-7*), followed by passive couplers of the type previously shown in *Figure 6-5*. 300-ohm to 75-ohm Balun transformers are used as required to match impedances. Active couplers — couplers with amplification — could be used. As shown in the dotted outline in *Figure 6-6*, the amplifier and coupler are combined into one active unit. The one shown in *Figure 6-8* couples to four TV sets through a 75-ohm cable. Other varieties have outputs for only two TV sets, a choice of 75- or 300-ohms impedances, and may even include special signal filters to trap out a particular interference.

► **Figure 6-6. A Simple Multiple-Set System Suitable for Home Use**

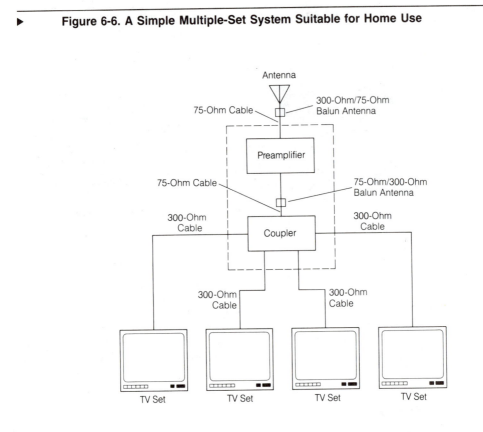

► **Figure 6-7. Booster Amplifier with 20dB Gain** *(Courtesy of Channel Master®)*

▶ **Figure 6-8. Active Coupler** *(Courtesy of Channel Master®)*

A much more elaborate general video distribution system is shown in *Figure 6-9*. This system combines the signals from an antenna, or cable TV system, a VCR, and up to one other signal source. The combined signals are then sent to various TV sets by coaxial cables.

Master Antenna Television System (MATV)

When many TV sets are to be operated from a single antenna, the system becomes much more complex than the simple systems described for the home. In recent years, systems such as the one illustrated in *Figure 6-10* have come into widespread use in apartment complexes, hotels, motels, hospitals and schools. An elaborate system such as this consists of three principle functions:

1. the antenna;
2. the distribution amplifier for amplification; and
3. the signal distribution system.

▶ **Figure 6-9. A Multiple-TV Home Video Distribution System** *(Courtesy of Radio Shack)*

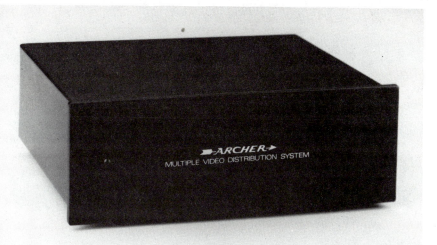

▶ **Figure 6-10. A Relatively Elaborate MATV System**

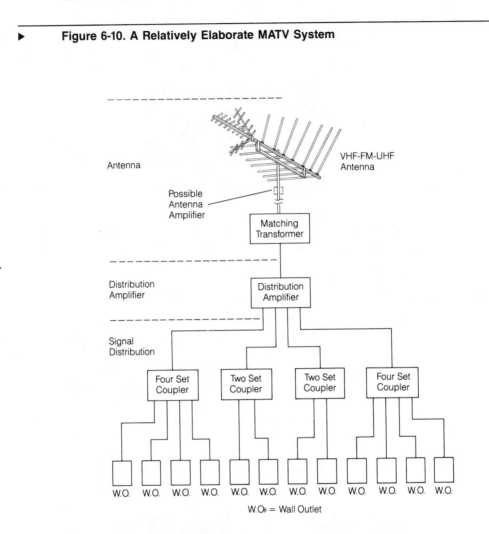

The Antenna

Choosing a MATV system antenna depends on several factors. These factors are similar to ones already discussed for other antennas. First, the location and direction to the transmitting antenna — can the receiving antenna be located to have line of sight reception, and what obstructions are in the way? Second, the distance to the transmitting antenna — is the receiving antenna in a strong signal area or a fringe area? Third, the receiving system complexity — is it a simple network or one that is very complex?

If the MATV system is in an urban area with high signal strength from local transmitters, a standard TV antenna may be all that is necessary. If in a low signal area, the system will require the same considerations discussed for a fringe area — a high gain, very directional antenna, maybe an antenna array with an antenna for each channel, and probably an antenna rotor. The signal from the antenna feeds the distribution amplifier (*Figure 6-10*), but individual mast-mounted preamplifiers may have to be used. Booster amplifiers that increase the signal from several antenna arrays also may be needed to overcome the transmission line losses to the distribution amplifier.

The Distribution Amplifier

The distribution amplifier boosts the signal to the power level necessary to provide signal for the many outlets. It also overcomes the losses in the transmission line, the couplers, isolation networks, and impedance matching circuits that make up the signal distribution section.

There are many types of distribution amplifiers with a variable combination of features. A simple 20 dB booster amplifier was shown in *Figure 6-7*, but an amplifier with more gain and/or additional features may be necessary. Some distribution amplifiers have more outlets (e.g. 4 to 8), and higher gain to accommodate additional outlets. A variety of signal controls may be available. A selector switch may determine whether FM signals are amplified or attenuated. There may be low band and high band VHF gain controls and a separate UHF gain control. It is not unusual for a distribution amplifier to amplify a received signal of 100 microvolts to several volts (up to 100 dB of gain) before distribution over the cable.

If separate channel antennas and amplifiers are used, they are combined before they are distributed so that all channels are available at each outlet.

The Signal Distribution Section

The MATV system is designed so that the signal strength at the receiver is about the same for each channel at each distribution point. This may require the use of intermediate active couplers in case the transmission lines to some of the distribution points are very long. Low-loss coaxial cable transmission lines are used for all external signal distribution.

In most internal installations, the transmission line inside the buildings is a 75-ohm coaxial cable rather than the familiar 300-ohm twin lead used in homes. The shielded coaxial cable is less susceptible to locally generated interference. In most cases, coaxial cable can be run in electrical power line conduits without excessive interference.

To insure sufficient isolation between receivers, a separate attenuator is used for each outlet. Also, this outlet may transform the 75-ohm impedance into a 300-ohm balanced impedance for connection to a TV receiver antenna terminals. Such transformation is not needed with many of the TV receivers today because they have 75-ohm impedance inputs.

SUMMARY

Problems in weak signal areas and their solutions have been covered in detail, as well as some tips for providing a multiple TV set distribution system. In the next chapter, we will concern ourselves with eliminating some of the noise and interference signals when receiving TV and FM signals.

TV and FM Noise and Interference

Throughout this book two subjects continue to be discussed — signal and noise. Signal is classified as the desired or wanted information in a transmission, and noise is classified as the unwanted information. We've looked at the wanted signal received by antennas in strong signal areas and in weak or "fringe" areas. In this chapter we will concentrate further on the unwanted signals. Generally, the unwanted signal is classified as noise; however, we will use the term noise for unwanted signals from natural sources, and the term interference for unwanted signals from man-made sources.

FM VERSUS AM

We have stated previously that FM receivers have much less noise than AM receivers. The reason is that most noise and interference amplitude modulate the signal carrier, and FM systems are designed to eliminate the unwanted signals that amplitude modulate the carrier. Since our purpose in this chapter is to find ways to eliminate noise and interference in TV and FM signals, especially that which is received by antennas, let's look further at AM and FM to better understand why FM systems have less noise.

AM Modulation

Figure 7-1 summarizes AM transmission and reception. Input information varies the amplitude of a transmitted carrier. The carrier frequency is held constant. The transmitted signals induce a voltage in the receiving antenna, the receiver amplifies the signals and detects the amplitude variations in the signal, and reproduces the transmitted information at the receiver output. Note that any interference signals that vary the amplitude of the received carrier become a signal in the AM receiver output. It is important to note that in a TV transmission, video (picture) signals amplitude modulate the carrier.

▶   **Figure 7-1. AM Transmission and Reception**

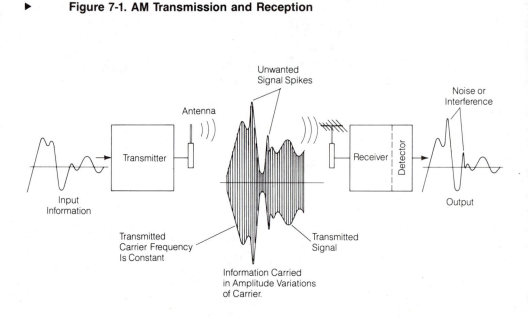

FM Modulation

Figure 7-2 summarizes FM transmission and reception. Input information varies the frequency of a transmitted carrier. The carrier amplitude is held constant. The transmitted signals induce a voltage in the receiving antenna, the receiver amplifies the signals, sends the signals through a limiter and discriminator, and reproduces the transmitted information at the receiver output. As shown, the limiter cuts off the tops and bottoms of the carrier signals to eliminate amplitude variations. Note the unwanted signals shown in *Figure 7-2*. They cause a variation in the received carrier amplitude at the receiving antenna. They do not appear in the receiver output because they did not vary the received carrier frequency. This is why FM transmission is essentially free of noise and interference that amplitude modulates the carrier.

TV sound is transmitted using frequency modulation. Video or picture information is transmitted using amplitude modulation. As a result, AM noise affects the TV picture, but the sound is free of much of the noise and interference that affects the TV picture. To know how to eliminate the unwanted signals, we need to look at the source of the signals.

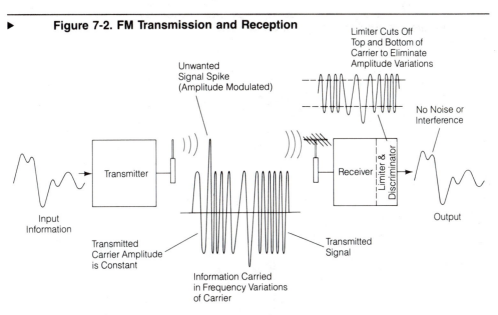

Figure 7-2. FM Transmission and Reception

SOURCES OF UNWANTED SIGNALS

As shown in *Figure 7-3*, unwanted signals can come from external sources or from internal sources. Internal sources usually are present whether there is a signal or not, and do not change abruptly unless something goes wrong inside the equipment or interconnections. External sources have two ways to be injected into the system of *Figure 7-3* — either through the antenna, or through the power input. The unwanted signals may or may not be present at all times. They may occur momentarily, intermittently or periodically. It is important when trying to eliminate the unwanted signals to know if they are getting into the system from external sources or whether they are present without any external input.

Sources of Noise

There are internal and external sources of noise. Two important internal sources are thermal noise and shot noise. An important external source is atmospheric noise.

Thermal Noise

All objects whose temperature is above absolute zero (0 degrees Kelvin) generate random electrical noise because of the vibration of the molecules within the object. This noise is called thermal noise. The noise power so generated depends only on the temperature of the object, and not on its composition. Because this is a fundamental property, noise is often defined by its equivalent noise temperature. The noise temperature of the air around us is approximately 300 degrees Kelvin (around 80 degrees F), while that of outer space is quite low (a few degrees K), and the sun is very high (around 5700 degrees K).

▶   **Figure 7-3. Sources of Unwanted Signals**

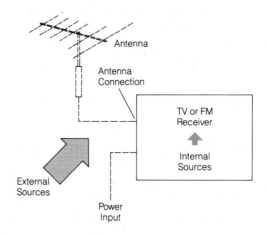

It is possible to build an amplifier whose equivalent noise temperature is well below its actual temperature, and thus it adds very little to the received noise. The low noise amplifier for the satellite system of Chapter 4 was classified in equivalent temperature to indicate its internal thermal noise.

Shot Noise

Temperature limited diodes, which includes virtually all semiconductor devices, generate shot noise when current is passed through the diode. The noise results because the current is carried by discrete particles (electrons and holes), and an impulse is generated by the passage of each particle. The noise is proportional to the current. The zero current noise is equal to thermal noise.

Atmospheric Noise

There is a noise that is intercepted by the antenna called atmospheric noise. It is present and would be the only signal if there were no other radiated signal. Atmospheric noise is very high at low frequencies, and decreases as frequency increases. It is present across the total broadcast band (AM radio) so that it is not eliminated with amplifier and antenna design; therefore, simple loop antennas are ordinarily adequate for AM reception since the noise cannot be filtered out. Atmospheric noise has decreased so much at TV and FM frequencies that a much better amplifier and antenna design is worthwhile to take full advantage of the lower noise. Atmospheric noise has fallen to nearly the thermal noise level at the upper VHF TV band, and stays at that level for all earth-bound communications until well into the upper microwave region, where it again begins to rise.

Signal-to-Noise Ratio

As the name implies, the signal-to-noise ratio is the ratio of two quantities — the signal power to the noise power. At any terminals, it is the ratio of the power of the signal that is present to the power of the noise that is present without any signal. Mathematically this is stated as follows:

$$\frac{S}{N} = \frac{Signal\ Power}{Noise\ Power}$$

The signal-to-noise ratio is usually expressed in dB. So,

$$\frac{S}{N}(dB) = 10\ \log\frac{S}{N}$$

Instead of power, signal and noise voltages can be measured. The expression for $\frac{S}{N}$ ratio in dB is different since power is proportional to the square of the voltage. It is

$$\frac{S}{N}(dB) = 20\ \log\frac{S}{N}\frac{(Signal\ Voltage)}{(Noise\ Voltage)}$$

Noise Figure

At a receiver, there will be separate signal-to-noise ratios at the input and the output. The noise figure of a receiver is defined as the ratio of the S/N at the receiver input (usually the antenna terminals) to the S/N at the receiver output. Noise figure is a measure of how much noise the receiver adds to the signal. The noise figure of a noiseless receiver is 1 (0 dB). This corresponds to a noise temperature of 0 degrees Kelvin. Using the above figures of merit, one can evaluate the receiver to determine how much noise it will contribute to the system. Receivers with high signal-to-noise ratios and low noise figures are the most desirable.

Improving Signal-to-Noise Ratio

Most of the methods to be discussed for reducing interference will also improve signal-to-noise ratio. The most effective of these include increasing the antenna height, changing the antenna orientation, using an antenna with higher gain and directivity, and using two or more antennas in combination. If the signal is very weak because the cable from the antenna to the receiver is very long or if the receiver gain is marginal, a preamplifier like the one described in Chapter 6 will increase the signal-to-noise ratio and improve receiver performance. Just a reminder, even though FM receivers are low noise, when the signal level approaches the noise level, FM receivers also fail, irrespective of whether the noise is internally or externally generated.

Sources of Interference

Interference basically is man-made except for atmospheric and weather conditions. The most noticeable being lightning discharges. Here are some examples:

1) automobile, truck and bus ignition,
2) electrical motors, arc welders, power lines,
3) neon and fluorescent lights,
4) some types of medical equipment (diathermy),
5) light dimmers,
6) computers,
7) other types of transmission, such as amateur radio, CB, aircraft, commercial communication, police and other public service radio, and even other FM or TV stations.

Generally sources that radiate intermittent or periodic signals are called impulse sources. Some examples are: electrical contacts that make and break, motor brushes that spark, flashing neon signs, solid-state light dimmers, and automobile ignitions. Lightning bursts are also included in this category. Impulse spikes are of short duration (microseconds) and often have amplitudes much greater than the signal being received. The interference may be either radiated as electromagnetic interference (EMI), or conducted on the power line, in the case of AC powered equipment.

Identifying Interference

One of the most important first steps in eliminating interference is to identify the type and source of the interference. Each type of interference that you see or hear on a receiver gives a clue to its identity by its appearance or sound (snow, ghosts, lines or bars, gibberish, etc.). A simple test may show whether it is internal or external to the set. External interference will usually disappear when the antenna is disconnected and the terminals are shorted. If it does, you know that the antenna is picking up the signal. If the interference stays, it is either a faulty receiver or it is coming in on the power line.

In general, there are three common cures for interference — eliminate the source; shield the antenna, power line input or total receiver; or filter the inputs to the receiver to trap the signals so they do not reach the receiver.

EMI

Electromagnetic interference can be identified by bands of dots of noise that sometimes move vertically on the TV screen and may cause distorted sound in the TV receiver or in an FM receiver. The most effective cure for EMI is to eliminate the source. As an example, popping sounds from the speaker or small flicks in a TV picture that change with changing engine speed and cease when the motor is turned off can be traced to an automobile ignition system. Resistor sparkplugs and resistance wiring will greatly reduce this type of interference.

The strength of a radiated EMI interfering field drops rapidly as the distance from the source of the interference increases. In most cases of EMI interference, if the interference continues on a regular basis, the source is close by.

When the EMI source cannot be found, removed, or relocated, then shielding or filtering are used to eliminate the interference. For example, the AC power line is a common way for EMI to get into a receiver. A filter of the type shown in *Figure 7-4a* can be inserted in the power line to filter out the unwanted signals. Or the EMI may be coming in through the antenna. A filter that attaches to the TV antenna terminals and is placed in line with the 300-ohm antenna lead-in is shown in *Figure 7-4b*. EMI from appliances and nearby neon lights can be eliminated or reduced significantly with such a filter.

When the noise source cannot be removed, then the antenna may have to be relocated so that a wall with metal lath or other shielding is between the antenna and the source. Or a new antenna might help. A relatively high, horizontally polarized antenna, with balanced or shielded lead-in is the least susceptible to this type of EMI.

Even with a properly placed antenna, the unwanted signal could be coupled into the receiver by the antenna lead-in. Open (unshielded) twin lead-in is a common cause. Changing to a shielded 300-ohm lead-in or to coaxial cable is likely to clear up such interference.

▶ **Figure 7-4. Appliance and Neon Interference Filters** *(Courtesy of Radio Shack)*

a. AC Line Interference Filter   b. 300-Ohm Antenna In-Line

RFI

Radio frequency interference (RFI) usually results in a series of straight or wavy lines or bars on the TV screen. A very strong RFI signal may cause complete loss of picture. It also may affect the audio system producing garbled sound, often unintelligible (sometimes pure gibberish). Most of this interference comes into the receiver through the antenna. Many times this type of interference is due to ham (amateur) radio or to Citizen's Band (CB) radio. High pass filters of the type shown in *Figure 7-5* are easy to install and can be very effective. The one shown (RS catalog no. 15-579) matches 75-ohm coax. It is installed in line with the antenna lead-in.

If filters do not work, try to eliminate the interference at the source. First, try to locate the source. A survey of the neighborhood for CB or ham radio antennas, industrial installations, etc., is a good start. The time of day and duration of the interference may also given a clue as to its origin.

After locating the source, if attempts to remedy the situation are unsuccessful, a written complaint to the nearest field office of the FCC (Federal Communications Commission) may be your only recourse.

Figure 7-5. RFI Filters *(Courtesy of Radio Shack)*

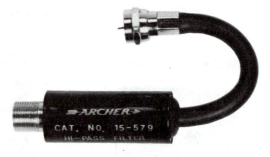

75-Ohm

MULTIPLE PATH SIGNALS

As described in Chapter 1 and Chapter 2, signals can arrive at the receiver through multiple paths, either radiated directly or reflected. Ghost images due to multiple path transmission may be reduced or eliminated by using the directional properties of the antenna. While watching the TV image, rotate the antenna for the best picture. In fixed installations this will usually require two people, one at the antenna and one at the set.

Reflections

Annoying reflections can occur as described in Chapter 2, but they also arise in long transmission lines when the impedance of the transmission line differs considerably from that of the antenna or the receiver input. Energy will be reflected from the receiver to the antenna and back to the receiver again. When the reflection reaches the receiver, a ghost is produced.

A ghost due to reflections on the line may be verified by "tuning" the line with hand capacity; that is, running the hand up and down the lead-in while watching the picture for change. This type of ghost can sometimes be reduced on a single channel by wrapping a 1 inch wide piece of aluminum foil around the lead-in, and sliding it up and down to find the best reception. This does not work on coaxial or shielded cable. For such cables, examine the system inter-connections and determine if the mismatch can be eliminated.

Overload

Sometimes too strong a signal can cause a problem called overload. The interference caused by overload may be reduced by inserting the attenuator, shown in *Figure 7-6,* in the antenna line at the set.

▶ **Figure 7-6. Signal Overload Attenuator** *(Courtesy of Radio Shack)*

Fading

When a received signal varies in intensity over a relatively short period of time, the effect is known as fading — one of the most troublesome problems encountered in radio communications. Fading may occur anytime a signal arrives at a point by two different paths — direct and skywave, reflected from an airplane, etc. When the signals combine, their phase may vary. Anytime they are out of phase (180 degrees is the most pronounced), then they cancel each other and fading results. A higher gain, more directional antenna will help a fading condition.

OTHER TYPES OF INTERFERENCE

Co-Channel Interference

Co-channel interference is particularly annoying in metropolitan areas where the stations are assigned frequencies very close together. In those areas where channel congestion exists, the nuisance effects may be minimized (if the stations are in different directions) by using a rotor to orient the antenna for minimum interference.

Capture Effect

FM systems exhibit a phenomenon called "capture effect," whereby the stronger of the two adjacent signals suppresses the weaker one. When trying to tune to the weaker station, the set will jump and lock on to the stronger station. Reducing the amplitude of the stronger signal will affect the amount of capture. About the only way to change the capture effect is to move or rotate the antenna, or obtain a more directional antenna and point it at the weaker station.

SUMMARY

In this chapter we've discussed the elimination of unwanted signals, and found that they can enter the receiver by other inputs than just the antenna. The best way to eliminate such signals is to remove the source. If that isn't possible, shielding or filtering is recommended. In the next chapter, antennas for mobile and cellular communications will be discussed.

CB and Cellular Antennas

Citizen band (CB) and cellular telephone systems are land mobile radio services. They are of most interest to the average consumer because the FCC (Federal Communications Commission) permits their use for personal and business communications. In this chapter, we will consider some basic concepts and antenna principles of both of these systems.

GENERAL CONSIDERATIONS OF LAND MOBILE RADIO

One segment of land mobile radio systems consists of mobile units and one or more base stations. The term "mobile station" is defined by the FCC as a station "used while in motion or during halts at unspecified points." In common practice, mobile station means a radio equipped car, truck, motorcycle, etc. A base station is permanently installed at a fixed location, and communicates with the mobile units and other base stations.

Another segment consists totally of handheld units — called walkie-talkies. In this segment, all units are mobile, and in most cases are tailored to specific applications — police, fire, disaster crews, toys, etc. The transmitter, receiver, and antenna are an integral part of the handheld unit. Even though some handheld units supply interchangeable telescoping or flexible antennas, there is no real choice of an antenna to be made.

CB and cellular systems have distinct differences. CB usually is operated independently of other communications systems; while the cellular system is an integral part of extending the common carrier telephone system.

SYSTEM OPERATIONAL MODES

Radio systems use three modes of operation — simplex, half-duplex and full-duplex. Land mobile systems use two modes — half-duplex and full-duplex.

Simplex operation is the transfer of information in one direction only. Broadcast radio stations and personal pagers use this mode. It requires only one frequency. A half-duplex system involves communication in two directions but not at the same time; that is, communication takes place in one direction then in the other. CB radios use the half-duplex mode. It requires only one frequency. In a full-duplex system shown in *Figure 8-1*, transmission and reception occurs in both directions at the same time. A and B both transmit and receive at the same time on different frequencies. Virtually all telephone systems, including cellular systems, use full-duplex operation.

▶ **Figure 8-1. A Full Duplex Radio System**

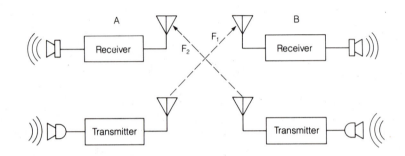

CB RADIO SYSTEMS

The communicating range between CB radio-equipped vehicles is usually much shorter than between vehicles and base stations or between two base stations. Mobile-to-mobile range usually is from 3 to 10 miles, base-to-mobile from 5 to 15 miles, and base-to-base range from 10 to 50 miles. The FCC rules prohibit CB communications over a greater distance than 150 miles. Because horizontally polarized antennas are bidirectional, base stations use vertically polarized omnidirectional antennas so communication can be maintained in all directions.

The FCC deliberately limits the range of CB systems by restricting base station antenna height to 20 feet above the natural formation, or existing man-made structures, or to 60 feet above the ground. The legal output power of the transmitter is limited to 4 watts. Some base stations increase their effective radiated power (ERP) in a given direction by using a directional antenna. If a base station is to be near an airport, check with the FCC for restrictions.

CB ANTENNAS

Many types of antennas are available for CB mobile and base station use. If the other system components are operating properly, the antenna is the most important part of a CB radio system, even more important than power output of the transmitter. Maximum range can be assured only if the antenna is as efficient as possible.

CB Mobile Antennas (Simple)

The length of a half-wave dipole at a CB frequency of 27 MHz is 17 feet. A quarter-wave vertical antenna would be 8.5 feet long. If mounted on a vehicle, the other 8.5 foot quarter wavelength is reflected in the body of the vehicle. A quarter-wavelength antenna may be considered as 100% efficient. Loading coils are added to antennas to better match them to the transmitter. A base-loaded antenna (loading coil at its base) has a comparative efficiency of about 80%. Similar efficiency is obtained with a center loaded antenna. The loading coils have high Q; that is, they absorb very little energy themselves resulting in more energy being radiated from the quarter-wave "whip" antenna. As a result, this type of antenna, whether base loaded or center loaded, is one of the most popular for mobile use. The efficiency drops rapidly as antenna length is shortened. Even so, there are many lengths and varieties of whip antennas for roof, trunk, fender, and mirror mounting.

Mounting the Mobile Antenna

CB antennas are mounted by many different methods and in many different locations. The most efficient and omnidirectional is a quarter-wavelength antenna mounted in the center of the vehicle's roof. *Figure 8-2a* shows a "shorty" roof-mount antenna. This "best" mounting is usually not practical because of the antenna's height; or the vehicle owner may not want the hole in the roof because of the effect on the vehicle's resale value. It is not practical to mount an antenna on the center of the roof of some vehicles — convertibles, vinyl roofs, etc.

▶ **Figure 8-2. CB Antenna Mountings** *(Courtesy of Radio Shack)*

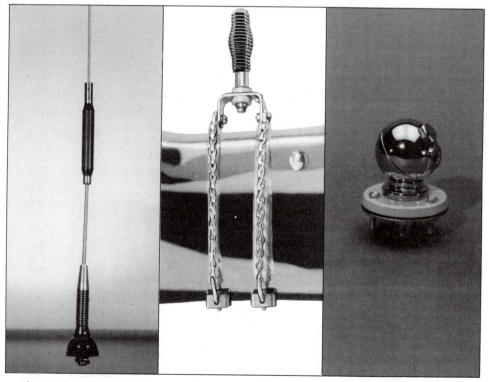

a. "Shorty" Roof-Mount Antenna   b. Dual-Chain Mount   c. "Split-Ball" Base

One good way to mount a quarter-wave whip is on the bumper of a car using clamps as shown in *Figure 8-2b*. To keep the antenna from breaking when it strikes an obstruction, a spring is used at the base as shown. The spring allows the antenna to bend over almost horizontally and then return to the upright position without damage. Because an antenna may be mounted in a variety of places, different sloping angles may be required of the base. A convenient "split-ball" mount (*Figure 8-2c*) can be turned and clamped in almost any position for such mounting.

The quarter-wave antenna also can be mounted on the rear deck just behind the body of the car or on the top side of a rear fender. Some units can be mounted without drilling holes. *Figure 8-3a* is an antenna that mounts with set screws. There are other units that have special mountings that snap in place.

The most efficient SWR will be obtained if the antenna feed point is above the effective ground plane of the vehicle.

▶ **Figure 8-3. Special Mounted CB Antennas** *(Courtesy of Radio Shack)*

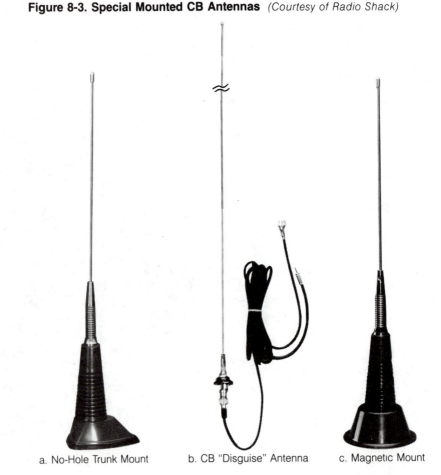

a. No-Hole Trunk Mount   b. CB "Disguise" Antenna   c. Magnetic Mount

A CB antenna on a vehicle is an advertisement to a thief that a transceiver is inside. If you hide the antenna, you hide the existence of the transceiver. One way to hide the antenna is to use the CB/AM/FM "disguise" antenna shown in *Figure 8-3b*. This combination antenna mounts in the same hole as a standard car radio antenna and is a tapered 32-inch stainless steel whip. The lead-in cable has a splitter and connectors for a CB transceiver and a car radio.

A simpler mounting uses a magnet-mount antenna like the one shown in *Figure 8-3c*. When in use, the antenna is held firm to the roof or trunk lid with a built in magnet; when not in use, it is removed and stored in the vehicle. Another idea is to use a motorized CB antenna that is extended to its full length when the CB is turned on and retracted when the CB is turned off. Still another idea — there are now transceivers designed to use the existing vehicle AM/FM radio antenna also for the CB. The transceivers have a tuning adjustment to impedance match the unit to the car radio antenna.

"Trucker" CB Antennas

You may have noticed the small antennas mounted on the cab mirrors of a tractor-trailer eighteen wheeler. They are CB antennas. They may be single antennas as seen in *Figure 8-4a* or "twin truckers" like the ones in *Figure 8-4b*. The two antennas, one on each side mirror, are about nine feet apart. They are called twin truckers because they are fed in-phase signals from a dual-phasing coaxial harness. The reception pattern is made more directional to the front and rear (down the road) with less pattern to the sides. This results in an effective increase in the range of the transceiver and eliminates some fading.

▶ **Figure 8-4. "Trucker" Antennas** *(Courtesy of Radio Shack)*

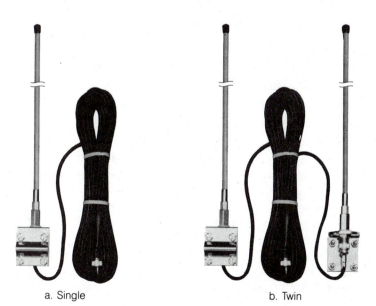

a. Single   b. Twin

CB Base Station Antennas

The effectiveness of the base station is dependent on the type and location of the antenna. Since the antenna system is common to both the transmitter and the receiver, the antenna characteristics effect both transmission and reception. The FCC limits the power output to 4 watts. This places additional emphasis on choosing the antenna carefully to obtain maximum performance from the system, and provide good system characteristics (high signal-to-noise, no fading, good clarity, etc.).

Choosing the Base Location

A base station antenna should be located as close to the transceiver as possible to reduce power loss in the transmission line. It should be mounted so it is not shielded by nearby objects. Since the effective radiation at 27 MHz is line of sight transmission, the height of the base station antenna above the average terrain is a very important factor. An example is shown in *Figure 8-5*. Antenna B is mounted much higher than antenna A, and has added range over Antenna A. At Point C, the signal from Antenna B is much stronger and more reliable, while the signal from Antenna A would probably be unusable.

▶   **Figure 8-5. Added Range Due to Increased Antenna Height**

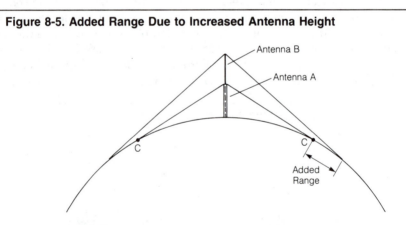

Radiation Pattern - Vertical Plane

Some additional range of the base station transmitter may be obtained by using a higher-gain antenna. In some installations, this alternative is used instead of increasing the height of the mast. A higher-gain antenna concentrates the energy of the radiation pattern horizontally instead of skyward. Skyward radiation is useless. The vertical plane radiation pattern is shown in *Figure 8-6*. The side view shows how the energy is concentrated horizontally in a pattern like a half of a donut. You can see how this pattern increases the transmitting range because the energy is kept close to the ground rather than going out into space. To summarize, a high-gain base station antenna should be located at the maximum legal and practical height to obtain the highest gain possible.

▶ **Figure 8-6. High-Gain Antenna Vertical Plane Radiation Pattern**

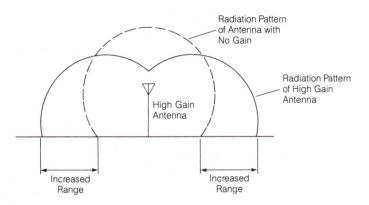

CELLULAR SYSTEMS

Even though mobile telephones have been available for many years, their use was limited. Previously, the quality of communications was poor, and only a small number of people were able to use them. A major breakthrough occurred when Bell Laboratories implemented CMRS, Cellular Mobile Radio Telephone Service or System. This started a mobile telephone revolution. With the cellular system, mobile telephone service may include all the features and services of a regular telephone system. In the future, a portable cellular telephone may be as common as a watch, pen or pocket calculator.

Basic Operation and Concept

The early mobile telephone system used one centrally located set of high power transmitters to communicate with all mobile units in the service area. Each of the system channels handled one call at a time.

CMRS is different. An area is divided into "cells" or units up to 24 miles in diameter with each having its own low-power transmitter, separate receiver, and control system. This combination of equipment is called the "cell site." The system is shown in *Figure 8-7*. The transmitter power is just strong enough to reliably reach any mobile unit within its borders, but too weak to interfere with communication in other cells on the same frequency. There are up to 832 assigned radio channels in the 850 MHz band. Different channels are assigned to neighboring cell sites to avoid interference between them. The channel frequencies may be reused at other remote cells in the service area. In a large urban area, the same channel frequency may be reassigned and simultaneously used 20 or more times. This principle of "frequency reuse" is the heart of CMRS. The cells are interconnected by a central computer called the Mobile Telecommunications Switching Office (MTSO) or Network Control Switch (NCS). This computer can simultaneously monitor the signal strength from thousands of cellular telephones. As a cellular telephone user moves from one cell to another,

▶   **Figure 8-7. A Cellular Telephone System**

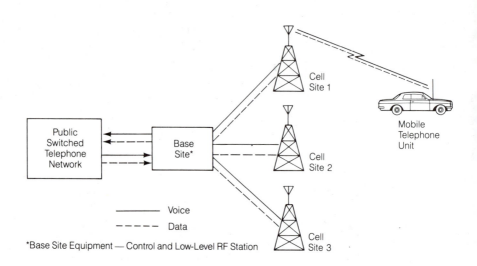

the telephone transmission is shifted to the next cell. This "handoff" is performed automatically by the computer at just the right instant. The handoff is so rapid that there is no interruption of communication.

Cell Site Antennas

At the cell site, each radio channel interfaces with the mobile unit by means of three antennas — one transmit antenna per 16 radio channels and two receive antennas. The two receiving antennas can be used in two different modes, either configured for omnidirectional coverage or for directional coverage. The cell site transmitting antennas are high gain, most generally omnidirectional, and vertically polarized. They are end-supported but are electrically center-fed to minimize antenna pattern distortion over the frequency band. They are approximately 13 feet long including the mount.

Cellular Telephone Antennas

Two things are necessary to achieve clear, noise-free telephone communications. The first is to obtain a cellular mobile antenna of good design, the next is to install it properly. An important feature is that the omnidirectional radiation pattern for the transmitted signal should be symmetrical. Whether the radiation pattern is symmetrical depends on the type and location of the antenna on the vehicle.

Cellular mobile antennas are made in several configurations. They may be one-quarter or five-eights wavelength, and some function without the "ground plane" effect used for CB antennas.

Various Mountings

Roof location gives the most omnidirectional coverage, but as with CB, either the customer or the construction of some vehicles does not permit an antenna and the associated transmission lines to be neatly installed in the roof. Alternate mounting locations on a vehicle are shown in *Figure 8-8.* By far, the most popular celluar telephone antenna is the "on-glass" type shown. The antenna is mounted with a special adhesive on the rear window or windshield. The radio signal is capacitively coupled through the glass using similar coupling boxes on each side of the glass.

Another popular cellular antenna is shown in *Figure 8-9* and is called the "elevated feed-point" antenna. It provides excellent performance mounted on the rear deck or fender of a car. Like the "on-glass" style, this antenna is designed to function without a ground plane. The radiation point of the signal is elevated above the vehicle to achieve an omnidirectional pattern.

▶ **Figure 8-8. Cellular Antenna Mountings** *(Courtesy of Radio Shack)*

a. Individual Antenna   b. Rear Window Mount

▶   **Figure 8-9. Elevated Feed-Point Antenna** *(Courtesy of Radio Shack)*

Portable Telephones

The mobile telephones we have considered thus far have been of the type mounted in a car or truck. Portable cellular telephones are now offered by many electronics stores. With a battery pack and flexible antenna, the telephone can go anywhere with you. Other portable cellular telephones are mounted in a briefcase. As miniaturization continues, soon there may be a telephone to go into your coat packet or on your wrist. It will be interesting to see what the antenna engineers will design to handle the innovative cellular system designs in the near future.

SUMMARY

Antenna characteristics are very important to the performance of the mobile CB radio and cellular telephone systems discussed in this chapter. Even though theory predicts such characteristics, much practical knowledge of antennas and their use come from amateur radio operators. Their "shortwave" antennas are the subject of the next chapter.

Shortwave Antennas

This chapter is about antennas used by amateur radio operators. They are classified under the title of shortwave antennas. Such antennas are designed using all the principles discussed in previous chapters; the difference is that they span a very wide range of frequencies.

THE RADIO AMATEUR

We've all heard the term "ham" radio operator. Or we have seen an unusual series of numbers and letters and the words "amateur radio operator" on automobile license plates. Both names—ham and amateur—identify a person whose hobby is experimenting with radio. Amateur radio operators have been around since the birth of radio communications nearly a century ago.

In the early years, the amateurs were assigned all of the unused frequency spectrum, which at that time was anything above 1500 kilohertz (1.5 MHz). Recall that wavelength can be calculated from the following equation:

$$\text{Wavelength (meters)} = \frac{300,000,000 \text{ (meters/sec)}}{\text{Frequency (cycles/sec)}}$$

If we substitute 1.5 MHz for the frequency as shown, then

$$\text{Wavelength} = \frac{300,000,000}{1,500,000} = 200 \text{ meters}$$

The wavelength is 200 meters. As a result, all the frequencies above 1.5 MHz were called (and still are today) "short waves" because their wavelength was shorter than 200 meters.

The early amateurs using "short waves" achieved unheard of long distant communications. Commercial communications, seeing this success, began using "short waves," and started crowding out the amateurs. To settle the issue, an international conference allocated the frequencies to be used for various activities, and reserved specific bands of frequencies for use by amateurs.

Amateur Frequency Bands

Figure 9-1 shows the frequency spectrum indicating the frequency allocation to the various well known services, and indicating the common amateur frequency bands. The lowest amateur frequency is the 160-meter band. This band, just above the AM broadcast band, is subject to special frequency and power limitations which vary according to geographical location. The limitations have been established to prevent interference with LORAN (long-range navigation) stations. *Table 9-1* indicates the major amateur bands by the common wavelength in meters. You will hear hams say that they are working at 15 meters or 10 meters, or 2 meters. This means they are transmitting and receiving on a frequency within the respective band.

▶ **Figure 9-1. Frequency Spectrum Showing Amateur (HAM) Bands**

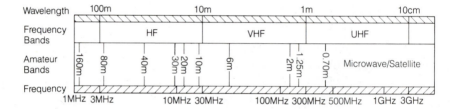

SHF Band — 3GHz to 30 GHz
EHF Band — 30GHz to 300 GHz

▶ **Table 9-1. Common Amateur Bands**

Band	Frequency (MHz)
160 meters	1.80 to 2.00
80 meters	3.50 to 4.00
40 meters	7.00 to 7.30
30 meters	10.1 to 10.15
20 meters	14.00 to 14.35
15 meters	21.00 to 21.50
10 meters	28.00 to 29.70
6 meters	50.00 to 54.00
2 meters	144.00 to 148.00
1.25 meters	222.00 to 225.00
0.75 meters	420.00 to 450.00

Besides the bands in *Table 9-1*, there are some 12 bands in UHF, SHF and EHF from 900 MHz to 300 GHz. Recently, conferences in Switzerland have resulted in further reallocation of certain portions of the frequency spectrum. If you desire more information, refer to *FCC Rules and Regulations*, Volumes II and VI, which are available with updated corrections in many libraries, or can be obtained from the Superintendent of Government Documents, U.S. Government Printing Office, Washington, DC.

TYPES OF SHORTWAVE ANTENNAS

Many different types of antennas are used by hams depending on the band of frequencies in which they are operating. No matter what amateur band is considered, the basic antenna theory discussed in the previous chapters still applies. For example, the free space radiation pattern and other characteristics of a half-wave 3 GHz dipole is the same as a half-wave dipole at 3 MHz. Only the size is different. The 3 megahertz dipole is 165 feet long, whereas the 3 gigahertz dipole is about 2 inches long!

Electrical Length vs Physical Length

Because it gives the half-wave length in feet, a somewhat easier equation for the half-wavelength of an antennas is:

$$\text{Half-wave Length (in feet)} = \frac{492}{\text{Frequency (in Megahertz)}}$$

However, the half-wave length that one actually uses for the physical length is about 5% shorter than that calculated with the above equation. The reason for this is that the electrical length of the antenna is different than its physical length. The antenna wire is usually fastened to insulators by making several twists or wraps at the ends. This produces what is known as "end effect." End effect, and the fact that energy travels along the antenna wire at slightly less than its free space velocity, makes the effective electrical length about 5% longer than the physical length.

Long Wire Antennas

Probably the most common type of antenna used by early radio amateurs was a long wire antenna. This type of antenna is a horizontal wire antenna which is at least one wavelength long, sometimes much longer. It may or may not have a terminating resistor, as shown in *Figure 9-2*. When no terminating resistor is used, it is bidirectional, with the maximum response off both ends. If the terminating resistor is used, maximum response is off the terminated end.

Zepp (or Zeppelin)

A type of end-fed antenna shown in *Figure 9-3* is called a Zepp antenna. They get their name because they were first used on the German lighter-than-air dirigibles called Zeppelins. This antenna is one-half wavelength at the lowest frequency. The impedance at the end is much higher (e.g. 1000 ohms or more) than the impedance at the center of a resonant dipole (75 ohms). Usually widely spaced 2-wire transmission line is used to couple to the antenna, and a matching box must be used between the transmitter and the transmission line.

Shortwave Listening

For the person interested in antennas for shortwave listening, a long wire antenna that is portable and can be used for quick, temporary setups is a coiled unit that clips over a telescoping antenna and can be extended to 23 feet like a measuring tape.

▶ **Figure 9-2. A Long Wire Antenna**

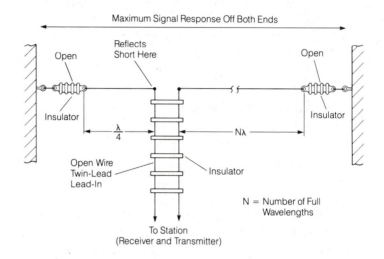

a. Without Termination

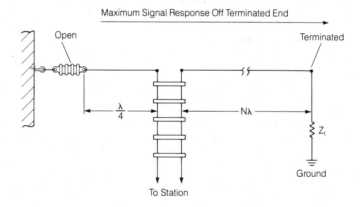

b. With Termination Z_t

Vertical Antennas

Another very popular amateur antenna is a vertical antenna. In Chapter 1 we showed the omnidirectional pattern of a vertical quarter-wavelength antenna. Generally, the radiation pattern of the vertical antenna is not deflected upwards by the ground nearly as much as the pattern of the horizontal antenna. *Figure 9-4* shows how increasing the length of the vertical antenna affects the radiation pattern. Increasing the frequency while keeping length constant produces exactly the same result.

► **Figure 9-3. End-Fed ZEPP Antenna**

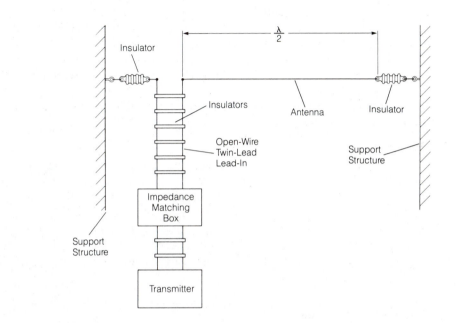

► **Figure 9-4. Radiation Pattern Due to Lengthening the Antenna**

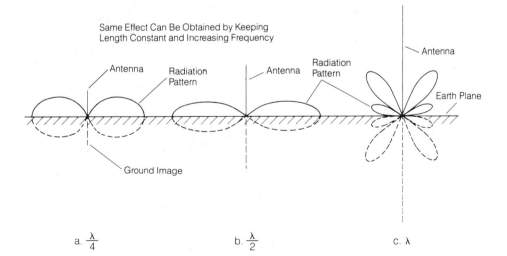

Loops

A loop consists of one or more turns of wire wrapped around a supporting structure that is shaped into a rectangle or a circle. *Figure 9-5* shows examples of loop antennas. A loop is quite directive and relatively compact, which makes it attractive for portable equipment. Its small size makes it easy to rotate for either maximum sensitivity or minimum interference. It is usually much smaller than the wavelength at which it is used. Its characteristics can be modified by changing the antenna Q (quality factor). This is usually done with a fixed or adjustable magnetic (ferrite) core. The radiation resistance of small loops is so low that it is nearly impossible to match a transmitter to them, so that loops are used primarily for receiving.

Beam Antennas

An antenna is called a beam antenna when its radiation/reception pattern is a very narrow beam to make it a very directional antenna. We've seen many examples of Yagi antennas which are beam antennas. Hams use beam antennas at all frequencies. You may have noticed very large Yagi type arrays mounted on a tower in the backyard of an amateur radio operator. In many cases, the antennas can be rotated. Beam antennas are about the only type shortwave antennas besides loops that can be rotated easily to permit efficient transmission and reception in any direction.

► **Figure 9-5. Loop Antennas**

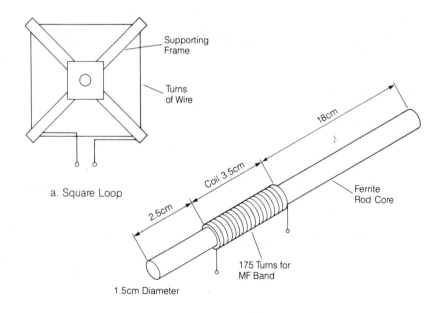

a. Square Loop

b. Ferrite (MF Band)

Quad-Cubical Antennas

An antenna that combines directivity and loop antenna characteristics is the quad-beam antenna shown in *Figure 9-6*. It is one of the more effective long range shortwave antennas. It consists of two or more square loops of wire supported by spreaders that were often made from bamboo, although fiberglass is more common today. The length of each side of the square is equal to a quarter wavelength of the transmitted signal. Loops may be either driven or parasitic. If parasitic elements are used, they have the same dimensions as the driven element. By tuning with shorted tuning stubs, the parasitic loop may be made to function as either a reflector or director, giving the antenna its directivity characteristics. In *Figure 9-6*, the parasitic element is a reflector.

Collinear Arrays

A collinear array consists of two or more elements mounted in-line along the same axis, with the current in each element in the same direction (in phase) as all other elements. The radiation pattern is bidirectional, with maximum radiation at right angles to the line of the elements. Although large collinear arrays may contain from 48 elements to as many as 128 elements, amateur arrays rarely contain more than 4 elements.

Other Antennas

There are other less frequently used types, such as the V and rhombic antennas, and broadside and endfire arrays. Information on these types, as well as more detailed descriptions of the antennas described above can be found in *The ARRL Antenna Handbook*, published by the American Radio Relay League, Newington, Connecticut, an association for radio amateurs.

▶ **Figure 9-6. A Two-Element Quad Antenna**

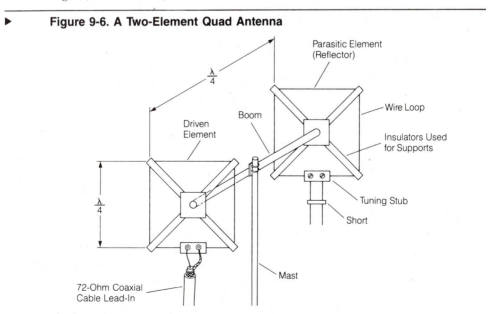

Antenna Placement

Most of the predictions of antenna performance are based on free-space behavior. Objects surrounding an antenna after it is mounted may affect its performance significantly. The erection of new buildings, the growth of trees, adjacent wires or towers are common environmental changes that occur to change an antenna's performance.

As we have stated many times throughout this book, one of the best ways to isolate an antenna is to mount it high above other obstructions. This may be very difficult for the very large and heavy antennas used for the lower frequency amateur bands. If the antenna is not mounted high enough its radiation/reception pattern will be affected. For example, *Figure 9-7* shows vertical radiation patterns of a horizontal antenna at various heights above ground. Generally the lower the antenna, the more the radiated (or received) wave is directed upwards. Consequently, the antenna position adversely affects long range communication. Also, the impedance of an antennas is dramatically reduced for heights much less than one quarter wavelength.

▶ **Figure 9-7. Horizontal Hertz Antenna Mounted at Various Heights**

All Views Are End Views

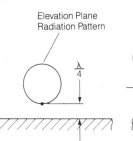
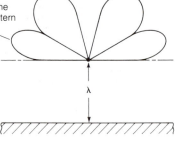

a. Quarter Wavelength　　b. Half Wavelength　　c. Full Wavelength

ANTENNA MEASUREMENTS

We have found that antenna performance characteristics vary because of a number of factors. Several that have been mentioned are: electrical length is different than physical length, the placement of the antenna, and the impedance matching of the antenna to its transmission line. It is necessary to have the correct characteristics for good signal reception when using any antenna, but it is particularly important if one is transmitting on a shortwave band. Maximum performance cannot be obtained unless the antenna and transmitter are properly adjusted. One of the best ways to verify that an antenna has the characteristics desired is to make several measurements after the antenna has been installed.

Grid-Dip Meters

One of the instruments that has been used for many years by ham operators to measure radio frequencies is the grid-dip meter. With this meter, it is possible to determine and adjust the frequency of tuned circuits, including antennas, without applying high power from the transmitter. A typical grid-dip meter is a handheld, battery-powered instrument containing a tunable oscillator with a scale or dial that is calibrated in frequency. The meter has a probe or coupling loop that couples energy into or out of the grid-dip oscillator (GDO) circuit. When the coupling loop is held near a tuned circuit and the frequency is varied, the resonant frequency of the tuned circuit can be detected. At resonance, the circuit absorbs the maximum power from the GDO. This causes a meter in the oscillator circuit to drop (dip) as the GDO is tuned through the resonant frequency of the circuit. In like fashion, when a loop is connected from the grid-dip meter to antenna terminals, energy from the oscillator is fed to the antenna and its resonant frequency may be determined.

In other applications, the GDO does not radiate energy but is used as an absorption wave meter to absorb energy from a "live" circuit. As the grid-dip meter is tuned to the frequency of the energy in the circuit, the meter rises to a peak. With a transmitter on and the grid-dip meter probe close to the antenna, the frequency of the transmitter is determined by tuning the grid-dip meter for a maximum reading. Or the grid-dip meter tuning can be set on the desired transmitter frequency, and the transmitter tuned until the meter reads maximum.

SWR

A very useful measurement to make on the transmission line from the transmitter to the antenna is the standing wave ratio (SWR). Recall that if there is a mismatch of impedances along the circuit including the transmitter, transmission line, and antenna, there will be an inefficient transfer of energy. Not all the energy will flow forward from transmitter to the antenna. Some of it will be reflected and set up a standing wave on the transmission line. A standing wave ratio meter is shown is *Figure 9-8.* It has cable connections so that it can be inserted into the transmission line and the standing wave ratio measured. By placing the meter selector switch in the CALIBRATION position, the forward energy measured by the meter is set at a calibration point (full-scale); in the SWR position, the meter measures the reflected energy and allows the SWR to be read directly. A properly matched circuit will have an SWR below 1.5 (about 4% reflected energy.)

To properly match the antenna to the transmission line, or vice versa, the meter is inserted between the transmission line and the antenna, and the antenna or transmission line impedance changed until the minimum reflected energy is obtained. This is a much easier and much more effective measurement than trying to measure the antenna impedance directly.

▶   **Figure 9-8. Standing Wave Ratio (SWR) Meter** *(Courtesy of Radio Shack)*

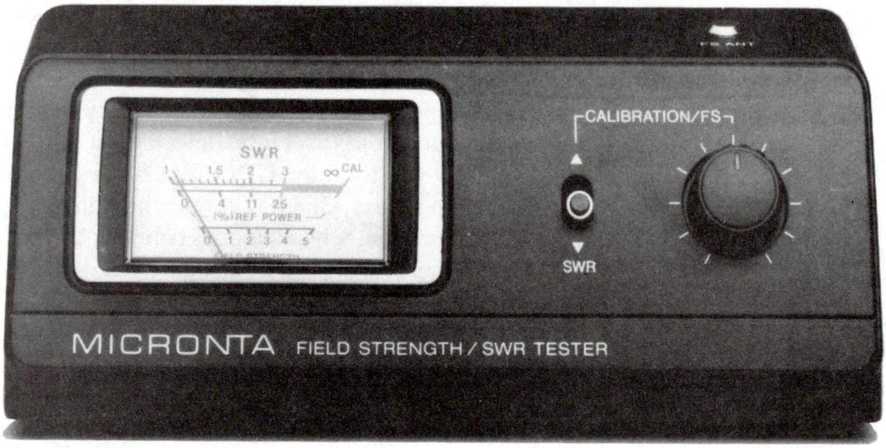

Field Strength

By putting an antenna probe on an SWR meter and placing the selector switch in the CALIBRATION/FS position, it can be used to measure relative field strength. The energy is picked up by the antenna, rectified, and read on the SWR meter. The meter readings will not be absolute readings but relative readings. But with such readings the radiation pattern of the antenna can be plotted.

SUMMARY

This chapter has concluded the discussion of antenna physical and electrical characteristics. To be effective, all antennas must be mounted and installed properly. The next chapter will deal with antenna installations, including safety precautions.

Antenna Installation

After you decide on an antenna and where it is going to be located for maximum performance, the antenna must be installed. Inside antennas present very few problems, but the installation of outside antennas for TV, CB and amateur radio must be planned very carefully. In this chapter we will consider the proper installation of antennas. Since hazards are involved, we will begin with a short refresher of some safety precautions that must be considered. A more complete consumer guide to video product safety is contained in the Appendix. The same precautions can be applied to all types of electrical and electronic equipment.

SAFETY

Each year serious injuries and even deaths occur because of accidents while installing an antenna, or from antennas or towers falling after installation. People fall off of ladders or from roofs; or they are electrocuted by contacting power lines. To help avoid some of the common accidents, here is a safety checklist to consider before beginning your installation.

1. Survey your planned installation carefully before you begin. Where are the power lines? Do you have proper clearance? Do you have good access to the place the antenna is going to be mounted? If you use guy wires, is there room for mounting them properly? Do you have adequate tools, and enough help?

2. If you are erecting a tower in a fringe area, or if the tower is to support a very heavy amateur radio antenna, make sure you have experienced help. If you need professional help, check the yellow pages under "Television Antenna Systems."

3. Power lines are particularly hazardous. Check the distance from the power line to the installation point. To be safe, the distance to any power line should be at least *twice* the length of the antenna assembly — the mast plus the antenna.

4. If you are going to be near overhead power lines at any point in the installation, don't use a metal ladder, and make sure no part of the antenna, mast, lead-in wire, or guy wires comes in contact with the power lines. These objects are all metal and will conduct electricity. It is a very dangerous situation. *IT CAN CAUSE A FATAL SHOCK.* Be very careful.

5. Don't do your installation on a windy day.
6. If you are going to be on the roof or in other high places, ask someone to be a spotter for you. Someone on the ground will be able to see things that you can't.
7. Assemble as much of the antenna and accessories on the ground as possible. Hoist it up, if you can, rather than carrying it up a ladder.
8. When using a ladder, remember the 1 for 4 rule (1 foot out for every 4 foot up). Anchor long ladders, climb them carefully, and have someone hold the ladder for you. The person holding the ladder should keep an eye out for items that might fall from above.
9. If during the installation an antenna starts to fall, get away from it and let it fall. That is much better than suffering serious injury yourself.
10. Check your local building codes for antenna installation specifications.
11. Again, plan the installation beforehand, make sure you have the proper tools and enough help, and work carefully.

CAUTION: In case of an accident involving a power line, the National Consumer Product Safety Council recommends the following:

1. If an antenna or any of its connecting parts touch a power line, DO NOT TOUCH the antenna or the power line. Do not remove it yourself. Call your local power company.
2. If a person has contacted a power line, or is touching an antenna that has contacted a power line, DO NOT TOUCH the person in contact with the antenna or power line — the electricity will travel on to you.
3. Use a DRY board, stick or rope to push or pull the antenna or power line away from the victim (or victim away from the antenna or power line, whichever is appropriate).
4. If the victim has stopped breathing, administer artificial respiration and stay with it.
5. Have someone call for medical help.

LIGHTNING AND LIGHTNING RODS

Once an outside antenna is installed, another hazard arises — protection against lightning. A rooftop antenna (or any high, exposed electrical conductor) may attract lightning during an electrical storm. If the antenna system is not properly protected, lightning damage to equipment and wiring can be severe, besides the potential for severe fire damage.

You probably have seen sharp metal posts at the highest point on farmhouses and barns. These are lightning rods. They are connected to ground rods and provide an easy highly conductive path for lightning to reach ground without going through the house or barn and causing damage. Antennas mounted in high places are just like lightning rods.

When lightning strikes an antenna, the current in the bolt must find a path to ground. As shown in *Figure 10-1*, there are two paths that the discharge current can take: 1. down the mast and 2. down the lead-in cable. It is very likely that the current will take both paths.

▶ **Figure 10-1. Two Paths for Lightning Discharge**

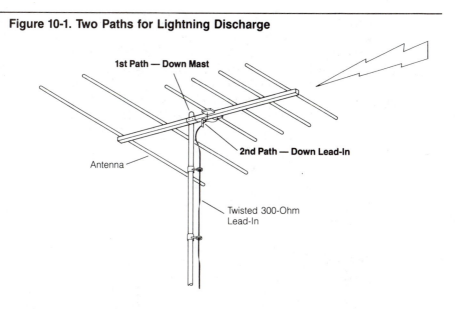

Grounding the Mast

The lightning discharge down the mast must has a good conductive path to ground. A copper wire of at least 10 AWG is recommended. Many local ordinances and codes prohibit the use of aluminum wire because of its poor conductivity and because it corrodes. If aluminum must be used, it should be at least 8 AWG. The ground wire must be connected securely to the base of the mast and run by the shortest path to ground. Do not make sharp bends or take complicated route. The lightning current will leave the wire and take another path, causing damage at unpredictable locations. The ground wire should be securely connected to a metal ground rod (Radio Shack UL listed Ground Rod, catalog no. 15-529). A typical installation is shown in *Figure 10-2*.

Static Discharge and Surge Absorbers

If the lightning discharge follows the lead-in wire, it may cause extensive damage to the equipment connected to the antenna. To avoid this, a low resistance path must be provided to ground before the lead-in cable enters the house. A static discharge device or coaxial surge absorber installed on the lead-in transmission line provides such a path. It can either be installed on the antenna mast (location #1 in *Figure 10-2)* where the grounded mast serves as the ground connection. Or it can be installed on the side of a building (location #2 in *Figure 10-2*) where a separate ground wire to the ground rod provides the ground path.

The static discharge device (*Figure 10-3*) has metal contacts that are separated by an air gap. The contacts are open. One contact is connected to the transmission line wire and the other contact is connected to the ground connection. When only RF signals are present on the transmission line, the air gap is open and is a high resistance to ground. When lightning strikes, a high voltage breaks down the air gap forming a current path and the current discharges to ground. Any static charge that accumulates on the antenna also discharges through the static discharge device or surge absorber. The static discharge or surge absorber is connected to the lead-in wire with the wing-nut solderless connectors shown in *Figure 10-3b*. The discharge unit and ground wire connections should be inspected annually. If it is detected that lightning has struck the antenna and discharged through the discharge unit, the discharge unit should be replaced. *Figure 10-3a* shows a coaxial unit.

▶ **Figure 10-2. Typical Installation**

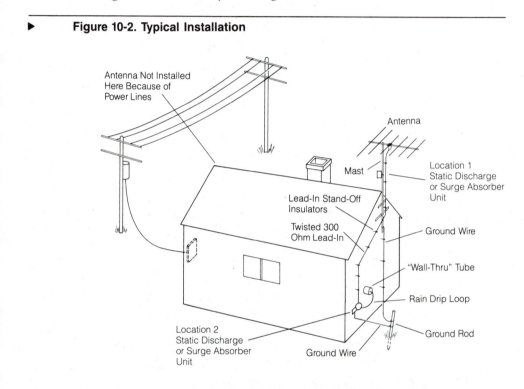

▶ **Figure 10-3. Static Discharge or Surge Absorber**

a. Coaxial Unit   b. 300-Ohm Unit

TV/FM ANTENNA INSTALLATION

Antennas require a mast, a base, and structures to support them. The support structure must be strong enough to support the weight of the antenna and the loading from high winds.

Selecting the Mounting Location

The site for your antenna should be selected first for safety, and then for performance. To determine a safe distance from power lines, telephone lines, trees and other obstructions:

1. Measure the length of the antenna mast or tower.
2. Measure the height of the antenna and add it to the length of your tower or mast.
3. Double the total for the minimum recommended safe distance from an obstruction.

If you do not have the required space, you should relocate the antenna. If the obstruction is a power line and the distance to the antenna is less than the minimum safe distance, you may want to consult a professional to see how the problem can be resolved.

Towers, Masts and Mounts

There are many types of support structures and mountings to choose from, including towers, telescopic masts, side-of-house mountings, and roof mounts.

Towers

The tower, shown in *Figure 10-4*, is used for TV antennas primarily in fringe areas, where the antenna must be higher than the roof of a house to obtain proper reception. Towers are usually triangular in shape, and are manufactured in sections that bolt together easily.

Heavy concrete footings are required to anchor towers. Unless a two-story house is available so that the tower can be erected close to the house and attached to it, guy wires will have to be used to stabilize towers. Climbing to the top of a tower and mounting an antenna is not something the normal layman can do. Consult with or hire professional help before you start your tower installation.

Masts

The telescoping antenna mast shown in *Figure 10-5* is made of interlocking sections which allow the antenna to be elevated for better TV and FM signal reception. Electronic Stores distribute this mast in two lengths, a 19-foot mast and a 36-foot mast. They are made of galvanized steel that interlock to prevent pull out, slip, or twist. Each ten foot section of these telescoping antenna masts should be guyed in three directions. The antenna can be mounted on the top section and the mast telescoped up as the guy wires are attached.

▶ **Figure 10-4. Antenna Tower (Used Mostly in Fringe Areas)**

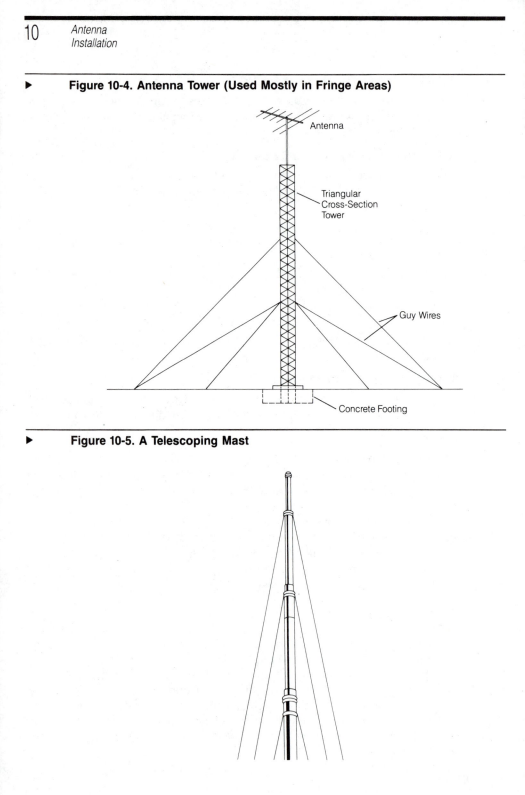

▶ **Figure 10-5. A Telescoping Mast**

House Mounts

Side-of-house mounts are shown in *Figure 10-6*. The one in *Figure 10-6a* is good for a brick house because the antenna is mounted on the wooden eave. A wall mount is shown in *Figure 10-6b*. This type is available in several sizes that stand off the mast from the wall 4 inches, 6 inches or 12 inches. An antenna using these mounts should not rise more than 10 feet above the rooftop. In addition, to prevent antenna collapse, make certain that adequate size lag screws are used to secure the mountings.

▶ **Figure 10-6. Side-of-House Mounts**

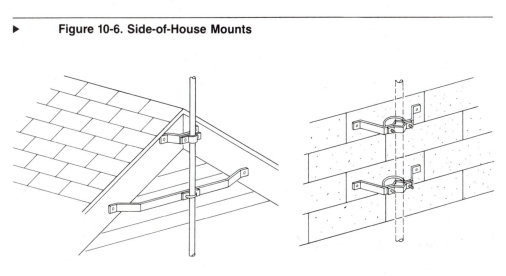

a. Eave Mount   b. Wall Mount

Chimney Mounts

The chimney is often an easy and convenient mounting place. Make certain that the chimney is strong and sound enough to support the antenna in high winds. If it has loose bricks or mortar, it should be repaired before using. Several types of chimney mounts are available for different sized chimneys. One type is shown in *Figure 10-7*. For maximum support, the mounting brackets should be placed as far apart as possible.

Roof Mounts

The roof mounts shown in *Figure 10-8* can be used on either peaked or flat roofs. *Figure 10-8a* is a base mount. It has a swivel foot that makes a convenient mount for many roof pitches. The tripod mount in *Figure 10-8b* is great for areas with high winds. It accommodates most roof slopes, and can accept masts up to 1¾" in diameter. Guy wires must be used for roof mounting if the mast is 6 feet or longer.

▶ **Figure 10-7. A Chimney Mount**

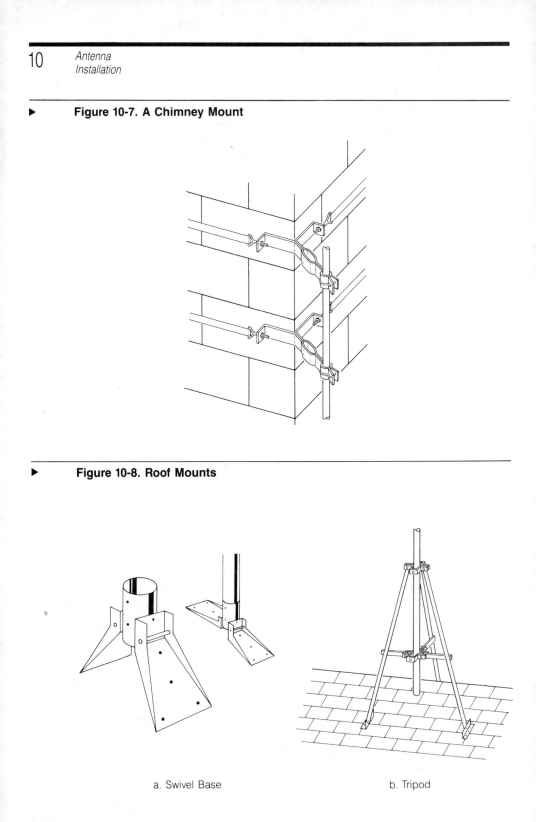

▶ **Figure 10-8. Roof Mounts**

a. Swivel Base   b. Tripod

Guy Wires

Guy wires should be used for tall antenna installations. They should be equally spaced in at least three directions around the mast and support the mast at least every 10 feet in mast height. A 45-degree angle between the ground (or support surface) and the wire is best for the guy wires, as illustrated in *Figure 10-9*. Guy wires must be anchored firmly. Anchors which screw into the ground are available at many building supply houses, and come in a variety of sizes. Alternatively, a hole may be dug in the ground with a posthole digger, and the bottom of the hole enlarged. The guy wire can be fastened securely around a short piece of pipe which is placed horizontally in the bottom of the hole, and concrete poured in the hole. The guy wires, brackets, rings, turn buckles, etc. are available from your local electronics store.

For tightening the guy wires, turn buckles are used in each guy wire (*Figure 10-9*). Tighten the lowest wires first, then the next higher ones, and so on, with the top guy wires being tightened last. Make sure prior to tightening that the mast is vertical (or plumb), and that each layer of guy wires are tightened equally to keep the mast aligned. If the top guy wires are tightened first, it is possible to bend the mast, sometimes causing it to break. Do not over tighten the guy wires.

► **Figure 10-9. Guy Wire Installation**

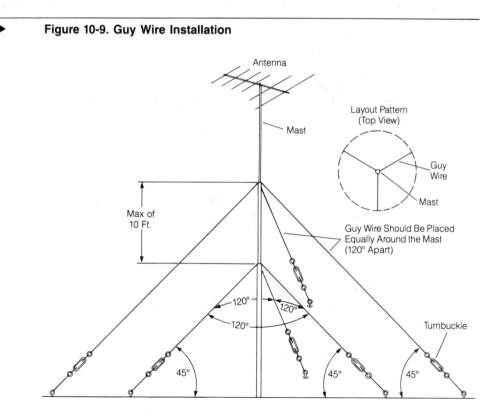

As the guy wires are tightened, the antenna is pulled down against its support. If you are not careful, a roof mounted antenna could be pulled right through the roof. Also, do not forget that guy wires can be a path for lightning discharge. Make sure the guy wires are grounded properly to the mast, especially the top ones, and that the mast is grounded.

Routing the Lead-In

When selecting the route for the transmission line from the antenna, avoid long, exposed horizontal runs which are likely to collect moisture or ice and dirt. Unshielded transmission lines (twin lead) should be kept at least 4 inches from metal objects. Capacitance between the metal and the transmission line alters the characteristics and causes attenuation or reflections in the line. Metallic or lead-based paint on unshielded twin lead can cause the same effect.

Twin-Lead Cable Installation

300-ohm twin lead is the most popular type of transmission line. To install twin lead, start at the antenna and thread it through the strain relief clamp mounted on the bottom of the antenna. Attach the twin lead to the terminal points tagged "connect lead-in here" with wing nuts and washers. Do not reuse old lead-in. Install new twin lead with any new antenna installation (or reinstallation), using sufficient stand-off insulators to prevent whipping in the wind. The stand-offs should be about four feet apart. Twist the lead-in at least two and a half times between stand-off insulators. Any interference induced in the lead-in will to be reduced because cancelling signals are being picked up by the twisted wires. Twisted lead-ins also take wind load better.

When the lead-in is run over the edge of the roof or around a corner, place the stand-off insulators so that the lead-in clears any metal by at least four inches. *Figure 10-2* illustrated a typical installation.

Some additional suggestions in routing the twin lead are:

1. Avoid aluminum storm doors and windows.
2. String lead-in so that it has minimum slack.
3. Avoid sharp bends in the lead-in.
4. Put a drip loop in the lead-in where it enters the house.
5. Bring the lead-in through the wall at a convenient location and use a wall outlet for a finished appearance.
6. DO NOT coil excess lead-in behind the TV set; cut it to length.

Coaxial Cable Installation

In some congested, high traffic areas, better results can be obtained with an RG-59/U coaxial cable transmission line. Coaxial cable does not require stand-off insulators. It can be taped to the mast, run along gutters or through aluminum windows. It also deteriorates less rapidly than twin lead; however, coaxial cable attenuates the signal more than twin lead. Coaxial cable requires cable connectors in order to correctly match the impedance of the transmission lines at connections. To match the impedance of a 300-ohm antenna to a 75-ohm coaxial cable transmission line, special Balun transformers are required. Also, such a transformer match may be required (in reverse) at the TV set if the TV set only has a 300-ohm antenna input. If the TV set has a 75-ohm antenna input, no transformer match is required. A schematic of the transmission line circuit is shown in *Figure 10-10*. As the lead-in comes into the house from the outside, whether it is 300-ohm twin lead or 75-ohm coaxial cable, special receptacles and wall plates are required to maintain the correct line impedance.

► **Figure 10-10. Transformer Matching Using Coaxial Cable from Antenna to TV Set**

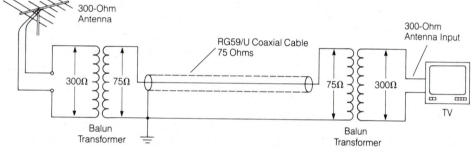

Wall Plates, Receptacles, Connectors

A handy way to get the lead-in through external walls (up to 13 inches thick) is the "Wall-Thru" lead-in tube. A special grommet seals out the weather. As shown in *Figure 10-2,* a rain drip loop in the transmission line just before it enters the tube keeps water from traveling along the line into the wall.

A variety of wallplates are available to give the lead-in installation a neat, professional appearance. These wallplates come with a wide assortment of connectors for both 300-ohm twin lead, 75-ohm coax, and for rotor control box cable. By using amplifiers, splitters, couplers and transformers, antenna connections may be provided at wall outlets in various rooms throughout the house.

There are a wide variety of in-line end connectors for both 300-ohm twin lead and 75-ohm coaxial cable. Many are the recommended "solderless" type. The 75-ohm coax connector used in most TV installations is called the "F" type. Radio Shack has these in many styles, including both screw-on and push-on types.

SUMMARY

This chapter concludes the book. Proper selection of an antenna, proper placement of an antenna, and proper installation of an antenna will assure maximum transmission and/or reception for a given application. Follow the guidelines outlined in this book and you should have success with your selection and installation.

Appendix

*A Consumer Guide to Product Safety

 This symbol is intended to alert the user of the presence of uninsulated "dangerous voltage" within the product's enclosure, that may be of sufficient magnitude to constitute a risk of electric shock to persons.

 This symbol is intended to alert the user of the presence of important operating and maintenance (servicing) instructions in the literature accompanying the appliance.

CAUTION

RISK OF ELECTRIC SHOCK
DO NOT OPEN

CAUTION

TO REDUCE THE RISK OF ELECTRIC SHOCK,
DO NOT REMOVE COVER (OR BACK).

NO USER-SERVICEABLE PARTS INSIDE.

REFER SERVICING
TO QUALIFIED PERSONNEL.

Video Products Important Safeguards

1. **Read Instructions** — All the safety and operating instructions should be read before the appliance is operated.

2. **Retain Instructions** — The safety and operating instructions should be retained for future reference.

3. **Heed Warnings** — All warnings on the appliance and in the operating instructions should be adhered to.

4. **Follow Instructions** — All operating and use instructions should be followed.

5. **Cleaning** — Unplug this video product from the wall outlet before cleaning. Do not use liquid cleaners or aerosol cleaners. Use a damp cloth for cleaning.

6. **Attachments** — Do not use attachments not recommended by the video product manufacturer as they may cause hazards.

7. **Water and Moisture** — Do not use this video product near water — for example, near a bath tub, wash bowl, kitchen sink, or laundry tub, in a wet basement, or near a swimming pool, and the like.

8. **Accessories** — Do not place this video product on an unstable cart, stand, tripod, bracket, or table. The video product may fall, causing serious injury to a child or adult, and serious damage to the appliance. Use only with a cart, stand, tripod, bracket, or table recommended by the manufaturer, or sold with the video product. Any mounting of the appliance should follow the manufacturer's instructions, and should use a mounting accessory recommended by the manufacturer.

9. **Ventilation** — Slots and openings in the cabinet are provided for ventilation and to ensure reliable operation of the video product and to protect it from overheating, and these openings must not be blocked by placing the video product on a bed, sofa, rug, or other similar surface. This video product should never be placed near or over a radiator or heat register. This video product should not be placed in a built-in installation such as a bookcase or rack unless proper ventilation is provided or the manufacturer's instructions have been adhered to.

*Courtesy of Radio Shack

10. **Power Sources** — This video product should be operated only from the type of power source indicated on the marking label. If you are not sure of the type of power supply to your home, consult your appliance dealer or local power company. For video products intended to operate from battery power, or other sources, refer to the operating instructions.

11. **Grounding or Polarization** — This video product is equipped with a polarized alternating-current line plug (a plug having one blade wider than the other). This plug will fit into the power outlet only one way. This is a safety feature. If you are unable to insert the plug fully into the outlet, try reversing the plug. If the plug should still fail to fit, contact your electrician to replace your obsolete outlet. Do not defeat the safety purpose of the polarized plug.

12. **Power-Cord Protection** — Power-supply cords should be routed so that they are not likely to be walked on or pinched by items placed upon or against them, paying particular attention to cords at plugs, convenience receptacles, and the point where they exit from the appliance.

13. **Lightning** — For added protection for this video product receiver during a lightning storm, or when it is left unattended and unused for long periods of time, unplug it from the wall outlet and disconnect the antenna or cable system. This will prevent damage to the video product due to lightning and power-line surges.

14. **Overloading** — Do not overload wall outlets and extension cords as this can result in a risk of fire or electric shock.

15. **Object and Liquid Entry** — Never push objects of any kind into this video product through openings as they may touch dangerous voltage points or short-out parts that could result in a fire or electric shock. Never spill liquid of any kind on the video product.

16. **Servicing** — Do not attempt to service this video product yourself as opening or removing covers may expose you to dangerous voltage or other hazards. Refer all servicing to qualified service personnel.

17. **Damage Requiring Service** — Unplug this video product from the wall outlet and refer servicing to qualified service personnel under the following conditions:

   a. When the power-supply cord or plug is damaged.

   b. If liquid has been spilled, or objects have fallen into the video product.

   c. If the video product has been exposed to rain or water.

   d. If the video product does not operate normally by following the operating instructions. Adjust only those controls that are covered by the operating instructions as an improper adjustment of other controls may result in damage and will often require extensive work by a qualified technician to restore the video product to its normal operation.

   e. If the video product has been dropped or the cabinet has been damaged.

   f. When the video product exhibits a distinct change in performance — this indicates a need for service.

18. **Replacement Parts** — When replacement parts are required, be sure the service technician has used replacement parts specified by the manufacturer or have the same characteristics as the original part. Unauthorized substitutions may result in fire, electric shock or other hazards.

19. **Safety Check** — Upon completion of any service or repairs to this video product, ask the service technician to perform safety checks to determine that the video product is in proper operating condition.

20. **Power Lines** — An outside antenna system should not be located in the vicinity of overhead power lines or other electric light or power circuits, or where it can fall into such power lines or circuits. When installing an outside antenna system, extreme care should be taken to keep from touching such power lines or circuits as contact with them might be fatal.

21. **Outdoor Antenna Grounding** — If an outside antenna or cable system is connected to the video product, be sure the antenna or cable system is grounded so as to provide some protection against voltage surges and built-up static charges. Section 810-21 of the National Electrical Code, ANSI/NFPA No. 70—1984, provides information with respect to proper grounding of the mast and supporting structure, grounding of the lead-in wire to an antenna discharge unit, size of grounding conductors, location of antenna-discharge unit, connection to grounding electrodes, and requirements for the grounding electrode. See Figure 1.

FIGURE 1   EXAMPLE OF ANTENNA GROUNDING AS PER NATIONAL ELECTRICAL CODE INSTRUCTIONS

[a]Use No. 10 AWG (5.3 mm²) copper, No. 8 AWG (8.4 mm²) aluminum, No. 17 AWG (1.0 m²) copper-clad steel or bronze wire, or larger, as ground wire.

[b]Secure antenna lead-in wire and ground wires to house with stand-off insulators spaced from 4 ft   (1.22 m) to 6 ft (1.83 m) apart.

[c]Mount antenna discharge unit as close as possible to where lead-in enters house.

[d]Use jumper wire not smaller than No. 6 AWG (13.3mm²) copper or the equivalent, when a separate antenna-grounding electrode is used. See NEC Section 810-21(j).

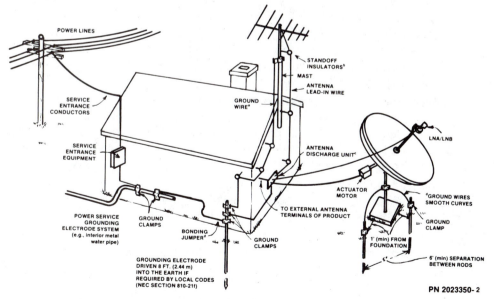

PN 2023350- 2

Glossary

Antenna preamplifier - A low-noise rf amplifier, usually mast-mounted near the terminals of the receiving antennas, used to compensate for transmission line loss and thereby improve the overall noise figure of the system.

Array - In an antenna, a group of elements arranged to provide the desired directional characteristics. These elements may be antennas, reflectors, directors, etc.

Balun - an acronym for BALanced to UNbalanced, also called balanced converter or "bazooka." A device used for matching an unbalanced coaxial transmission line to a balanced two-wire system.

Booster amplifier - A circuit used to increase the output current or the voltage capabilities of an antenna distribution amplifier circuit without loss of accuracy (ideally) or inversion of polarity.

Cassegrain antenna - An antenna the feed of which is positioned near the vertex of the reflector, with a small subreflector placed near the focal point. The feed illuminates the subreflector and the subreflector redirects the waves toward the main reflector, which then forms the radiated beam.

Characteristic impedance - The impedance of an infinite length of transmission line (twisted pair, twin-lead, coaxial cable, etc.). It is also the impedance of any length of line terminated by its characteristic impedance. It is uniquely determined by the size and spacing of the conductors and the properties of the dielectric (insulation).

Corner reflector - A reflecting object consisting of two (dihedral) or three (trihedral) mutually intersecting conducting surfaces. Trihedral reflectors are often used as radar targets. The surfaces may be solid, mesh, or a matrix of metal rods.

Decibel - The decibel is 10 times the logarithm of a power ratio. 1 dB is a power ratio of 1.259. 10 dB is a power ratio of 10.

Dipole antenna - Also called dipole. A straight radiator usually fed in the center. Maximum radiation is produced in the plane perpendicular to its axis. The length specified is the overall length.

Direct wave - A wave that is propagated directly through space as opposed to one that is reflected from the sky or ground.

Director - A parasitic antenna element located ahead of the driven element on the main antenna beam used to increase the radiation and directivity of the antenna in the direction of the major lobe.

Distribution amplifier - A power amplifier designed to increase signal power in an antenna distribution system. Its output impedance is sufficiently low that changes in the load do not appreciably affect the output voltage.

Driven element - An antenna element connected directly to the transmission line. It is the major radiating element of a transmitting antenna, and the major "excited" element of a receiving antenna.

Electric field - 1. The region about a charged body. Its intensity at any point is the force which would be exerted on a unit positive charge at that point. 2. The invisible lines of force that surround an antenna that has a voltage applied to it.

Electromagnetic wave - A wave of both electric and magnetic fields which travels in free space at the velocity of light. Radio waves, light, x-rays, and infrared rays are all examples of electromagnetic waves.

Field strength - The effective value of the electric field intensity in volts per meter produced at a point by an electromagnetic wave. Unless otherwise specified, the measurement is assumed to be in the direction of maximum field intensity.

Folded dipole antenna - An antenna comprising two parallel, closely spaced dipole antennas. Both are connected together at their ends, and one is fed at its center.

Frequency modulation (FM) - Modulation of a sine-wave carrier so that its instantaneous frequency differs from the carrier frequency by an amount proportionate to the instantaneous amplitude of the modulating wave. Combinations of phase and frequency modulation also are commonly referred to as frequency modulation.

Fringe area - The area just beyond the limits of the reliable service area of a television transmitter. Signals are weak and erratic, requiring the use of high-gain, directional receiving antennas and sensitive receivers for satisfactory reception.

Full-duplex transmission - Transmission and reception in 2 directions at the same time. It requires 2 frequencies. Cellular telephone uses full-duplex transmission.

Ghost - Also referred to as ghost image. An undesired duplicate image offset somewhat from the desired image as viewed on a television screen. It is due to a reflected signal traveling over a longer path and hence arriving later than the desired signal. It may be eliminated by the use of a directional antenna array which receives signals over only one path.

Grid-Dip Oscillator (GDO) - Simple test equipment for sensing a signal and identifying its frequency, or for adjusting a circuit's frequency or resonance.

Half-duplex transmission - Transmission of information in 2 directions, but not at the same time. CB radios use half-duplex transmission.

High-band VHF - Television channels 7 through 13, covering a frequency range of 174-216 MHz.

Impedance - A measurement of the opposition to current in a circuit or current element. It has two components, resistance and reactance, and is expressed in ohms. It is defined as the ratio of voltage to current at the measuring point. Its reciprocal is called admittance. Symbol: Z.

Indoor antenna - Any receiving antenna located inside a building but outside the receiver.

Interference - Any electrical or electromagnetic disturbance, phenomenon, signal or emission, largely man-made but also natural (lightning), which causes or can cause undesired response, malfunctioning, or degradation of the performance of electrical and electronic equipment.

Log-periodic antenna - A type of directional antenna that achieves its wideband properties by geometric iteration. The radiating elements and the spacing between elements have dimensions that increase logarithmically from one end of the array to the other. The impedance of the antenna varies periodically with frequency.

Low-band VHF - Television channels 2 through 6 covering frequencies between 54 and 88 MHz.

Magnetic field - An area where magnetic forces can be detected around a permanent magnet, natural magnet or any conductor carrying a current.

Microvolt - One millionth of a volt.

Millivolt - One thousandth of a volt.

Noise - Any unwanted disturbance in a system; e.g., undesired electromagnetic radiation in a transmission channel or device.

Omnidirectional - Also called nondirectional. All-directional; not favoring any one direction. Usually refers to the horizontal plane only. A truly omnidirectional antenna would be called isotropic, if it existed.

Parabolic reflector - Also called a reflector element. One or more conductors or conducting surfaces for reflecting radiant energy from a source into a directed beam.

Reflector - A parasitic antenna element located opposite the general direction of maximum transmission or reception.

Signal strength - The strength of the signal produced by a transmitter at a particular location. Usually it is expressed as so many millivolts per meter of the effective receiving antenna length.

Simplex transmission - Transmission of information in one direction. AM or FM radio is simplex transmission.

Stacked array - An antenna array consisting of two or more interconnected antennas, usually one above the other. This is done to increase the gain and horizontal directivity.

Standing wave - When an electromagnetic wave is sent down a transmission line which is not terminated with its characteristic impedance, some of the energy is reflected back from the termination. In some locations the reflected wave reinforces the direct wave, in others it opposes the direct wave. The result is a stationary pattern of waves along the line called standing waves.

Standing Wave Ratio (SWR) (also VSWR, Voltage Standing Wave Ratio) - The ratio of maximum to minimum voltage of a standing wave along a transmission line.

Television channel - A band of frequencies suitable for the transmission of television signals. The channel for associated sound signals is normally considered part of the television channel.

Transmission line - One or more conductors (wires) used for propagating electromagnetic energy from one place to another.

Transponder - A radio transmitter-receiver which automatically transmits a signal on the reception of a proper input signal.

Twin lead - Also called twin line. A type of transmission line covered by a solid insulation and comprising two parallel conductors, the impedance of which is determined by the diameter and spacing of the conductors and the insulation dielectric. The three most common impedance values are 75, 150, and 300 ohms.

Yagi antenna - An end-fire antenna that consists of a driven dipole (usually a folded dipole), a parasitic dipole reflector and one or more parasitic dipole directors usually all in the same plane.

Yagi-Uda antenna - A Yagi antenna with more than one driven element. Usually designed for use at FM frequencies.

Index

BUSINESS REPLY MAIL

FIRST CLASS MAIL   PERMIT NO. 1317   INDIANAPOLIS IN

POSTAGE WILL BE PAID BY ADDRESSEE

HOWARD W. SAMS & COMPANY

2647 WATERFRONT PKY EAST DR

INDIANAPOLIS IN 46209-1418

Dear Reader: *We'd like your views on the books we publish.*

PROMPT Publications, an imprint of Howard W. Sams & Company, is dedicated to bringing you timely and authoritative documentation and information you can use.

You can help us in our continuing effort to meet your information needs. Please take a few moments to answer the questions below. Your answers will help us serve you better in the future.

1. Where do you usually buy books?_____

2. Where did you buy this book?_____

3. Was the information useful?_____

4. What did you like most about the book?_____

5. What did you like least?_____

6. Is there any other information you'd like included?_____

7. In what subject areas would you like us to publish more books?

   (Please check the boxes next to your fields of interest.)

   □ Amateur Radio   □ Computer Software

   □ Antique Radio and TV   □ Electronics Concepts/Theory

   □ Audio Equipment Repair   □ Electronics Projects/Hobbies

   □ Camcorder Repair   □ Home Appliance Repair

   □ Computer Hardware   □ TV Repair

   □ Computer Programming   □ VCR Repair

8. Are there other subjects not covered in the checklist that you'd like to see books about?

   _____

9. Comments _____

   _____

Name_____

Address_____

City_____ State/Zip_____

Occupation _____ Daytime Phone_____

Thanks for helping us make our books better for all of our readers. Please drop this postage-paid card in the nearest mailbox.

For more information about PROMPT®Publications, see your
authorized Sams PHOTOFACT® distributor.
Or call 1-800-428-7267 and ask for Operator MP2.

Imprint of Howard W. Sams & Company
2647 Waterfront Parkway East Drive, Indianapolis, IN
46214-2041

61025